# Logical Design
# of Digital Systems

# DIGITAL SYSTEM DESIGN SERIES

**ARTHUR D. FRIEDMAN,** *Editor*
*University of Southern California*


BREUER    *Digital System Design Automation:*
*Languages, Simulation & Data Base*
BREUER and FRIEDMAN    *Diagnosis & Reliable Design of*
*Digital Systems*
vanCLEEMPUT    *Computer Aided Design of Digital Systems —*
*A Bibliography*
vanCLEEMPUT    *1975-76 Update Computer Aided Design of Digital*
*Systems — A Bibliography*
FRIEDMAN    *Logical Design of Digital Systems*
FRIEDMAN and MENON    *Theory & Design of Switching Circuits*

# Logical Design
# of Digital Systems

**ARTHUR D. FRIEDMAN**

*Department of Electrical Engineering
and Computer Science Program
University of Southern California*

**PITMAN**

PITMAN PUBLISHING LIMITED
39 Parker Street, London, WC2B 5PB

*Associated Companies*
Pitman Publishing Co. SA (Pty) Ltd, Johannesburg
Pitman Publishing New Zealand Ltd, Wellington
Pitman Publishing Pty Ltd, Melbourne

First Published in Great Britain 1977
First published in USA 1975

©1975 Computer Science Press, Inc.,
   4566 Poe Avenue, Woodland Hills, California 91364

Library of Congress Cataloging in Publication Data

Friedman, Arthur D.
   Logical design of digital systems.

   Bibliography: p.
   1. Switching theory.   2. Logic circuits.   3. Computers—Circuits.   I. Title.
TK7868.S9F74      621.3819′58′2      74–82932

US ISBN 0-914894-50-1
UK ISBN 0 273 01061 1

Printed by photolithography and bound in Great Britain by
Unwin Brothers Limited, The Gresham Press, Old Woking, Surrey.

# Contents

# Preface

The study of the techniques of digital circuit and system design have been referred to as *logic design* or *switching theory*. Although frequently used interchangeably, these terms may have different connotations. Thus logic design may be used to refer to the development of procedures for the design of digital systems. In practice many design techniques are heuristic in nature and may be based primarily on the experience and knowledge of the circuit designer. Switching theory studies properties of switching circuits and considers the development of design procedures which optimize some parameter of the design. Frequently concepts originally developed as theory become important aspects of practical design. Furthermore adaptation to technological change is simplified for the person with a strong theoretical background. Consequently I have attempted to cover both practical design techniques as well as theoretical problems with emphasis on general concepts and problems of potential relevance to current technologies.

The book is intended for an introductory one semester undergraduate or basic graduate level computer science or electrical engineering course in switching theory or logical design of digital systems. No specific previous knowledge is required on the part of the student other than some general mathematical ability. The book features many topics relevant to current technologies which are usually not presented in an introductory text. With the advent of integrated circuits and especially large scale integration, the classical switching theoretic design objective of component minimization has been replaced by an interest in system level design or modular design, in which systems are designed as interconnected sets of relatively complex modules. Related topics presented herein include modular design of combinational circuits, logical completeness, system level design using register transfer languages and logic design using ROM's and PLA's. Beginning from the simplest precepts of gate level logic design, the concepts of combinational circuit design, sequential circuit design, and finally system level design are developed in a systematic and unified manner, culminating in an introduction to the design of digital computer systems, which is presented as a natural extension of circuit design rather than a totally different subject.

The first two chapters introduce the subject of digital systems, and present basic required mathematical concepts, including set theory, Boolean algebra, number systems, arithmetic procedures, and codes. In Chapter 3 procedures for the design of minimal combinational circuits are developed. More advanced topics of combinational circuit design are covered in Chapter 4. Topics include multilevel circuit design by factorization which is illustrated by the design of decoders, modular realizations, illustrated by the design of a comparison circuit and a parallel adder, logical completeness, and special classes of combinational functions. A new category, fanout-free functions, of relevance to magnetic bubble technology and to the subject of easily diagnosable circuit design, is presented for the first time in any book. In Chapter 5 the subject of sequential circuit design is presented. Topics include sequential machines and state tables as models and representations of sequential circuits, the state assignment problem and state minimization of completely specified state tables. The subject of sequential circuit experiments is also briefly discussed. Chapter 6 develops a register transfer language and demonstrates how it can be used to design complex systems from basic combinational and sequential circuit modules, including the specification of a control unit as a sequential machine. A model of a stored programmed digital computer is presented and the concepts of macro and microinstructions are introduced. Finally logic design using ROM's and PLA's is considered. Chapter 7 introduces the problems of digital system physical design and digital system diagnosis and testing. This is intended primarily to provide insight into other problems related to digital system design, and add perspective to the role of logic design in the global design problem.

The basic ideas are presented in an informal manner and illustrative examples are used to demonstrate these concepts. In addition formal proofs of most theorems and algorithms are presented. Problems at the end of each chapter range from simple exercises to problems of considerable difficulty and some unsolved problems (demarcated by a flag (†)). (A solution manual is available for course adoptions). Each chapter also contains a guide to the literature for topics covered in that chapter. Although the complete contents of the book can be covered in a rapidly moving one semester course, it is possible to omit certain sections (indicated in the table of contents by a flag (†)) without loss of continuity.

I wish to express my gratitude to Professor S. H. Unger of Columbia University, New York, NY who initially stimulated my interest in this subject. I also wish to thank Mr. J. Goldstein for drawing the figures. Finally to my wife Barbara, for her encouragement and assistance.

Arthur D. Friedman

# part 1

# Basic Principles

# CHAPTER 1

# *Fundamental Concepts*

## 1.1 INTRODUCTION

Digital switching circuits are characterized by the feature that signals (voltages and currents) in the circuit are restricted to a finite set of possible values. Such circuits have many important applications. In a telephone system digital circuits are used to connect a call. The inputs to the system are a sequence of dialed digits which specify the desired connection. There are thus a finite number of inputs and a finite number of possible connections and hence the network which creates the connection can be digital. Digital systems are also useful in pattern (character) recognition machines. If a character is written on a fine grid, each section of the grid can be interpreted as being black or white. These grid segments can thus be converted to electrical signals with two possible values. The character can then be decoded by a digital system. Undoubtedly the most important application of digital circuits is in digital computers. These systems can perform sequences of arithmetic computations at very high speeds under the control of a program stored within the system. We shall consider these systems in detail in Chapter 6.

In most digital circuits, signals are restricted to two possible values. In this book we shall be concerned with the analysis and synthesis of such circuits and their interconnection to form digital systems. Digital circuits are studied at two different levels of detail. In one level, interest is focused on analysis and synthesis of these circuits in terms of basic electrical components such as transistors, diodes, resistors, etc., and also voltage levels, currents, wave shapes, etc. In this book, we will

3

be concerned with digital circuits in terms of their *logical behavior*. We consider circuits composed of elements called *gates* whose inputs and outputs are constrained to have only two possible values. Such circuits are called logic circuits. Once a circuit has been designed at the logic circuit level, the corresponding physical circuit can be obtained by suitable interconnections of subcircuits implementing the gates used in the logic circuit. In addition to the implementation of the logic circuit, problems related to timing, wave shapes, loading, etc., have to be solved during circuit design. We shall not consider these problems in this book.

The subject of designing digital circuits for a prescribed logical behavior has been called both *logic design* and *switching theory*. These terms have frequently been used interchangeably although they may have different connotations. Logic design tends to emphasize development of procedures for the design of switching circuits, whereas switching theory studies properties of switching circuits and attempts to discover design procedures which optimize some parameter of the design. Obviously, these disciplines are closely related and frequently concepts originally developed as theory become important aspects of practical design.

Another related subject is *automata theory*, which considers the capabilities and limitations of various mathematical models of computational processes. We will include those results which enhance our understanding of the capabilities of actual digital circuits.

As mentioned earlier, we shall be primarily concerned with the analysis and design of digital circuits whose basic components are gates.* Gates are designed from connections of basic electrical components such as resistors, diodes, and transistors. Some simple gates designed from diodes and resistors are shown in Figure 1.1.

In normal switching circuit applications diodes are operated so that the voltage on the anode side cannot exceed the voltage on the cathode side.† Hence, if the input voltage levels $e_1$ and $e_2$ in the gate of Figure 1.1(a) are constrained to the two possible values $V_+$ and $V_-$ where $V_+ > V_-$, then if $V_H > V_+$, the output voltage $e_o$ is approximately equal to the minimum $(e_1, e_2)$ (i.e. the minimum of $e_1$ and $e_2$). If $e_o >$ minimum $(e_1, e_2)$ then for one of the diodes the anode side voltage

---

*Previous generations of digital circuits were frequently designed from relays and contacts. The operation of such circuits is described in the Appendix.

†Actually the anode side voltage level may exceed the cathode side voltage level but only by a very small amount.

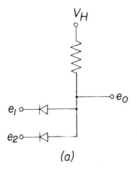

(a)

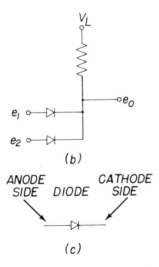

(b)

ANODE                    CATHODE
 SIDE    DIODE            SIDE

(c)

Figure 1.1   Basic diode-gates

would be larger than the cathode side voltage which is not possible.

The table of Fig. 1.2(a) specifies the output voltage $e_o$ for each of the possible inputs on $e_1$ and $e_2$ for the gate of Figure 1.1(a). Thus row 3 of this table is interpreted as follows: If $e_1 = V_+$ and $e_2 = V_-$, then $e_o = V_-$. Instead of using $V_+$ and $V_-$ to represent the two possible voltage levels it is common to use the symbols 0 and 1. Although $V_+ > V_-$ it is not necessary to associate 0 with $V_-$ and 1 with $V_+$. There are two possible correspondences $(V_+ \rightarrow 1, V_- \rightarrow 0)$ or $(V_+$

| $e_1$ | $e_2$ | $e_o$ |
|-------|-------|-------|
| $V_-$ | $V_-$ | $V_-$ |
| $V_-$ | $V_+$ | $V_-$ |
| $V_+$ | $V_-$ | $V_-$ |
| $V_+$ | $V_+$ | $V_+$ |

(a)

| $e_1$ | $e_2$ | $e_o$ |
|-------|-------|-------|
| 0 | 0 | 0 |
| 0 | 1 | 0 |
| 1 | 0 | 0 |
| 1 | 1 | 1 |

(b)

| $e_1$ | $e_2$ | $e_o$ |
|-------|-------|-------|
| 1 | 1 | 1 |
| 1 | 0 | 1 |
| 0 | 1 | 1 |
| 0 | 0 | 0 |

(c)

Figure 1.2    First diode-gate table

$\rightarrow 0$, $V_- \rightarrow 1$). If the first interpretation is used, the resulting circuit is said to have *positive logic* while the second interpretation corresponds to *negative logic*. The tables of Figure 1.2(b) and (c) display the input/output relationship of this gate for positive logic and negative logic assumptions respectively.

The gate of Figure 1.1(b) can be analyzed in a similar manner. In this case if $V_L < V_-$ the output $e_o$ is equal to the maximum $(e_1, e_2)$. The input/output relationship is expressed in the table of Figure 1.3(a) in terms of the voltage levels $V_+$, $V_-$, and in the tables of Figures 1.3(b) and (c) for positive and negative logic assumptions respectively, in terms of the symbols 0 and 1.

The number of inputs on a gate can be increased by connecting addition-al diodes with the same orientation to the common junction point. In practice the output voltage of a gate may be slightly different from the two possible voltage levels associated with the gate inputs. This signal degradation problem is aggravated when gates are connected to

| $e_1$ | $e_2$ | $e_o$ |
|-------|-------|-------|
| $V_-$ | $V_-$ | $V_-$ |
| $V_-$ | $V_+$ | $V_+$ |
| $V_+$ | $V_-$ | $V_+$ |
| $V_+$ | $V_+$ | $V_+$ |

(a)

| $e_1$ | $e_2$ | $e_o$ |
|-------|-------|-------|
| 0 | 0 | 0 |
| 0 | 1 | 1 |
| 1 | 0 | 1 |
| 1 | 1 | 1 |

(b)

| $e_1$ | $e_2$ | $e_o$ |
|-------|-------|-------|
| 1 | 1 | 1 |
| 1 | 0 | 0 |
| 0 | 1 | 0 |
| 0 | 0 | 0 |

(c)

Figure 1.3    Second diode-gate table

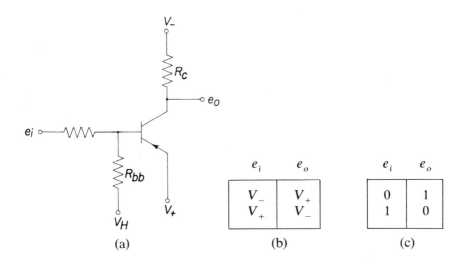

| $e_i$ | $e_o$ |
|---|---|
| $V_-$ | $V_+$ |
| $V_+$ | $V_-$ |

| $e_i$ | $e_o$ |
|---|---|
| 0 | 1 |
| 1 | 0 |

(a)                    (b)                    (c)

Figure 1.4    A transistor gate

form complex circuits. By using amplification at appropriate points in the circuit, this problem can be eliminated. The transistor circuit of Figure 1.4(a) can be used to provide such amplification if the resistor values are properly selected. The input/output voltage level relationships are specified in the table of Figure 1.4(b) and in terms of the symbols 0 and 1, for both positive and negative logic, by the table of Figure 1.4(c). The diode gates and transistor circuit can be combined to form individual gates as shown in Figures 1.5 and 1.6.

Similar gates can be designed from other basic electrical elements. The logic designer usually is not concerned with the precise manner in which a gate is designed or with the choice between positive or negative logic. These decisions may be governed by requirements of compatibility with other subsystems, power supply consideration or many other factors which are not of interest in the design of circuits on the logic level. In this process, the designer works in terms of the signal value symbols 0 and 1 and gates whose operation can be described by tables describing the input/output relationship in terms of these signal values, rather than actual voltage levels and actual physical devices which realize gates. One such gate type is the AND gate. The output of a 2-input AND gate will be 1 if and only if both of the inputs are 1. We will represent the signals on a gate by variables of the form

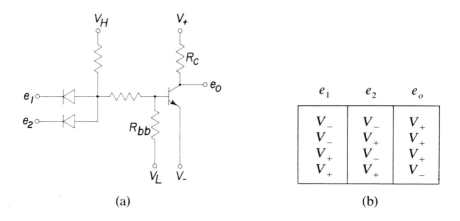

| $e_1$ | $e_2$ | $e_o$ |
|-------|-------|-------|
| $V_-$ | $V_-$ | $V_+$ |
| $V_-$ | $V_+$ | $V_+$ |
| $V_+$ | $V_-$ | $V_+$ |
| $V_+$ | $V_+$ | $V_-$ |

(a)                              (b)

**Figure 1.5   A diode-transistor gate and its logical behavior**

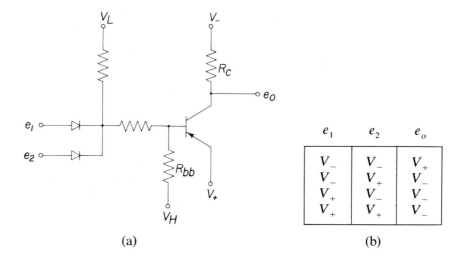

| $e_1$ | $e_2$ | $e_o$ |
|-------|-------|-------|
| $V_-$ | $V_-$ | $V_+$ |
| $V_-$ | $V_+$ | $V_-$ |
| $V_+$ | $V_-$ | $V_-$ |
| $V_+$ | $V_+$ | $V_-$ |

(a)                              (b)

**Figure 1.6   Another diode-transistor gate and its logical behavior**

$x_i$ (for gate inputs) and $z_i$ (for gate outputs). A 2-input AND gate will be represented as shown in Figure 1.7 and its behavior is described by the accompanying table. The AND gate can be generalized to have more than two inputs. The output of an n-input AND gate is 1 if and only if all $n$ inputs are 1.

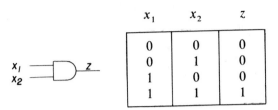

| $x_1$ | $x_2$ | $z$ |
|-------|-------|-----|
| 0 | 0 | 0 |
| 0 | 1 | 0 |
| 1 | 0 | 0 |
| 1 | 1 | 1 |

Figure 1.7   An AND gate

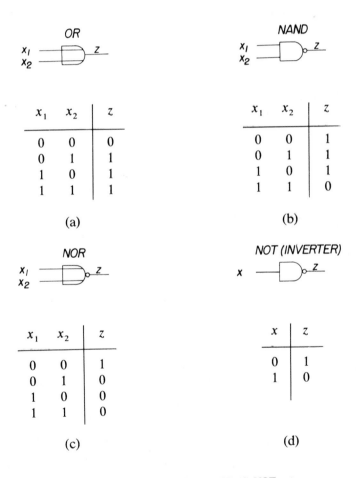

OR

| $x_1$ | $x_2$ | $z$ |
|-------|-------|-----|
| 0 | 0 | 0 |
| 0 | 1 | 1 |
| 1 | 0 | 1 |
| 1 | 1 | 1 |

(a)

NAND

| $x_1$ | $x_2$ | $z$ |
|-------|-------|-----|
| 0 | 0 | 1 |
| 0 | 1 | 1 |
| 1 | 0 | 1 |
| 1 | 1 | 0 |

(b)

NOR

| $x_1$ | $x_2$ | $z$ |
|-------|-------|-----|
| 0 | 0 | 1 |
| 0 | 1 | 0 |
| 1 | 0 | 0 |
| 1 | 1 | 0 |

(c)

NOT (INVERTER)

| $x$ | $z$ |
|-----|-----|
| 0 | 1 |
| 1 | 0 |

(d)

Figure 1.8   (a) OR (b) NAND (c) NOR (d) NOT gates

The other basic gate types we will be using are the OR, NAND, NOR, and NOT (INVERTER) gates. The behavior of these gates and their representations are shown in Figure 1.8. The OR gate output is 1 if and only if at least one of its inputs are 1. The NAND output is 1 if and only if at least one of its inputs is 0. The NOR output is 1 if and only if all of its inputs are 0. All of these gates can have more than 2 inputs. The NOT gate is constrained to have only one input and the output is 1 if and only if the input is 0. From the tables of Figure 1.2 we conclude that the circuit of Figure 1.1(a) can be used as an AND gate for positive logic or as an OR gate for negative logic. Similarly, the gate of Figure 1.1(b) is a positive logic OR and a negative logic AND. The transistor circuit of Figure 1.4 is a NOT gate. The circuit of Figure 1.5 is a positive logic NAND and a negative logic NOR and the circuit of Figure 1.6 is a positive logic NOR and a negative logic NAND. Note that a NAND is formed by combining a *NOT* with an *AND* while a NOR is formed by combining a *NOT* with an *OR*.

In this book we shall study how these basic elements can be combined to design digital systems which can perform useful functions. However, we must first consider some important mathematical concepts which are essential to the study of digital circuits.

## 1.2 MATHEMATICAL BACKGROUND

### 1.2.1 SET THEORY

Of fundamental importance in the study of switching circuits is the mathematical concept of a set. A *set* is simply a collection of objects which are referred to as the *elements* of the set. Sets are commonly denoted by capital letters $A,B,C$, etc., (possibly subscripted) and the elements of a set are denoted by small subscripted letters $a_1,a_2,a_3....$ If an object $a_i$ is an element of set $A$ we will say that $a_i$ is a member of $A$ or is contained in $A$ and denote this fact by $a_i \epsilon A$. If a set $A$ has a finite number of elements $a_1,a_2...a_n$ we can represent the set by listing its elements within brackets $A = \{a_1,a_2...a_n\}$ and say that $A$ is a *finite set*. The order of listing of the elements in the representation of a set is irrelevant. Thus the set containing elements $a_1,a_2,a_3$ could be represented as $\{a_1,a_2,a_3\}$ or $\{a_3,a_1,a_2\}$, etc. The fact that an element $b_1$ is not a member of $A$ will be denoted by $b_1 \notin A$. The elements in a set $A$ can also be specified by some characteristic property of all the elements of $A$ and only those elements. For example, we could speak of the set of even positive integers. We represent the statement that a set $A$ consists of all elements $x$ such that $x$ has property $P$ as $A = \{x \mid x$ has property $P\}$. The number of elements in a finite set $A = \{a_1, a_2 \ldots a_n\}$ will be denoted by $|A|$. In this case $|A| = n$. One important set is the set which does not contain any elements. This set is denoted by $\emptyset$, which is called the null set and $|\emptyset| = 0$. Another important set is $I$, the identity set which contains all elements from some implicit universal set of elements. If we denote this universal set as $S$, $I = \{x \mid x \in S\}$.

Sets can be combined in such a way as to create new sets. The *union* of two sets, $A,B$, denoted by $A \cup B$, is a set which contains as members all elements of $A$ and all elements of $B$. Thus if $A = \{1,2,3\}$ and $B = \{3,4,1\}$, $A \cup B = \{1,2,3,4\}$. Since $A \cup B = B \cup A$ and $(A \cup B) \cup C = A \cup (B \cup C)$, this operation can unambiguously be extended to multi-set unions $A_1 \cup A_2 \cup \ldots \cup A_n$ which we denote by $\cup_{i=1}^{n} A_i$. This set contains all elements which are contained in any of the sets $A_1, A_2, \ldots A_n$. The *intersection* of two sets $A,B$, denoted by $A \cap B$, is a set which contains all elements contained in $A$ which are also contained in $B$. Thus if $A = \{1,2,3\}$ and $B = \{3,4,1\}$, $A \cap B = \{1,3\}$. The multi-set intersection $A_1 \cap A_2 \cap \ldots \cap A_n$, denoted by $\cap_{i=1}^{n} A_i$, is a set which contains all elements which are contained in *all of the sets* $A_1, A_2, \ldots, A_n$. Given two sets $A$ and $B$, set $A$ is *contained* in $B$ (denoted $A \subseteq B$) if every element of $A$ is also an element of $B$. $A$ is also called a *subset* of $B$. Sets $A$ and $B$ are equal ($A = B$) if $A \subseteq B$ and $B \subseteq A$. Set $A$ is *properly contained* in $B$ ($A \subset B$) if every element of $A$ is also an element of $B$ and some element of $B$ is not an element of $A$, (i.e., $A \subseteq B$ and $B \not\subseteq A$). $A$ is then called a *proper subset* of $B$. For any set $A$, $\emptyset \subseteq A \subseteq I$. If $A \not\subseteq B$ and $B \not\subseteq A$, the sets $A$ and $B$ are *noncomparable*. For any set $A$, $A \cap I = A \cup \emptyset = A$, $A \cup I = I$, $A \cap \emptyset = \emptyset$.

## Example 1.1

Consider the sets $A = \{1,2,3\}$, $B = \{1,2,3,4\}$, $D = \{1,2,3,5\}$, and $E = \{4,5,6\}$. Then $A \subset B$ and $A \subset D$. However, $B \not\subseteq D$ and $D \not\subseteq B$. Therefore, $B$ and $D$ are noncomparable. $B \cap D = A$, $B \cup D = \{1,2,3,4,5\}$, $A \cap E = \emptyset$, $A \cup E = \{1,2,3,4,5,6\}$.

A set $A$ can be visually represented as a circle (as shown in Fig. 1.9(a)) where the area within the circle represents the elements in $A$, and the area outside the circle represents elements not in $A$ but in

$S$, the universal set. Two sets $A,B$, which have the property that $A \subset B$ could be represented as shown in Fig. 1.9(b), where the shaded area represents those elements contained in $B$ but not in $A$. Two sets $A$ and $B$ such that $A \not\subset B$, $B \not\subset A$ and $A \cap B = \emptyset$ are represented in Fig. 1.9(c). If $A \cap B \neq \emptyset$ these two sets are as represented in Fig. 1.9(d). The union of these two sets is represented by the shaded area within either circle. The intersection is represented by the crosshatched area common to both circles. The set of elements in $A$ but not in $B$, denoted by $A - B$, is represented by the shaded area in Fig. 1.9(e), and is referred to as the *difference* of $A$ and $B$. The set $I - A$ is denoted as $\bar{A}$ and is depicted in Figure 1.9(f).

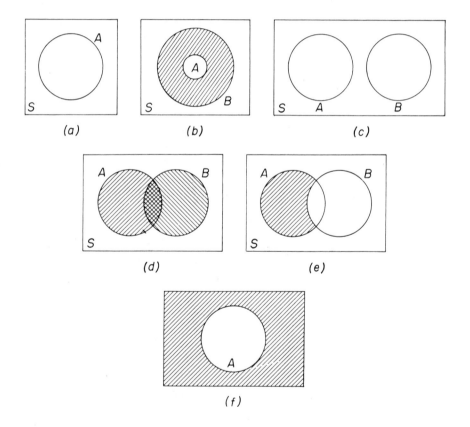

**Figure 1.9** Venn diagram representation of (a) a set $A$, (b) $A \subset B$, (c) $A \cap B = \phi$, (d) $A \cup B$ and $A \cap B$, (e) $A\text{-}B$, (f) $\bar{A}$

Such representations are called *Venn diagrams* and are useful for informal proofs of set theoretic results as illustrated in the following example.

**Example 1.2**

We will demonstrate that $A \cup (B \cap C) = (A \cup B) \cap (A \cup C)$. First consider the expression on the left hand side of the equation. This set

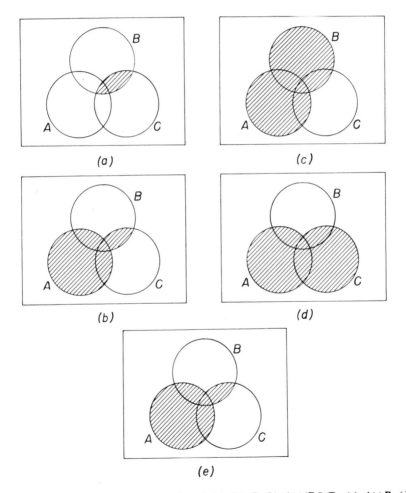

**Figure 1.10** **Venn diagram representation of** (a) $B \cap C$, (b) $A \cup (B \cap C)$, (c) $A \cup B$, (d) $A \cup C$, (e) $(A \cup B) \cap (A \cup C)$

can be derived by forming the intersection of $B$ and $C$ and taking the union of that set with $A$ as illustrated in Figures 1.10(a) and (b). Similarly, $(A \cup B) \cap (A \cup C)$ is formed as depicted in Figures 1.10(c), (d), (e), where the shaded area in Figure 1.10(e) is the area which is shaded in both (c) $(A \cup B)$ and (d) $(A \cup C)$. Since (b) and (e) have identical shaded areas, $A \cup (B \cap C) = (A \cup B) \cap (A \cup C)$.

## 1.2.2 RELATIONS

An important concept in mathematics is that of a *relation*. Some common numerical relations are *less than* (e.g., 2 is *less than* 4) and square (e.g., 4 is equal to the *square* of 2). Let us consider a binary relation which is defined between two elements. To formally define such a relation it is helpful to first define *ordered pairs* of elements. If $a$ and $b$ are objects belonging to sets $A$ and $B$ (not necessarily distinct) the *ordered pair* denoted by $(a,b)$ consists of $a$ and $b$ in that order. That is, $(a,b) = (c,d)$ if and only if $a = c$ and $b = d$ whereas $\{a,b\} = \{c,d\}$ if and only if $a = c$ and $b = d$ or $a = d$ and $b = c$.

The set of all ordered pairs $(a_i, b_j)$ where $a_i \epsilon A$ and $b_j \epsilon B$ is called the *Cartesian product of $A$ and $B$ and is denoted by $A \times B$. A relation $R$ from $A$ to $B$ is any subset of $A \times B$. If $(a,b) \epsilon R$ we say that $a$ is in relation $R$ to $b$ and denote this by $aRb$.

The set of all elements $a$ in $A$ such that $(a,b) \epsilon R$ for some $b$ in $B$ is called the *domain* of $R$ and the set of all elements $b \epsilon B$ such that $(a,b) \epsilon R$ for some $a \epsilon A$ is called the *range* of $R$.

**Example 1.3**

Let $A = \{1,2\}$, $B = \{1,2,3,4\}$. Then $A \times B = \{(1,1), (1,2), (1,3), (1,4), (2,1), (2,2), (2,3), (2,4)\}$. The relation $aRb$ if and only if $b$ is the square of $a$ defines the subset of $A \times B$, $R = \{(1,1), (2,4)\}$. The relation $aR'b$ if and only if $a < b$ defines the subset of $A \times B$, $R' = \{(1,2), (1,3), (1,4), (2,3), (2,4)\}$.

If a relation $R$ is defined from $A$ to $B$ and $A = B$, then $R$ is said to be *on $A$*. There are several important properties associated with binary relations on a set $A$. Such a relation $R$ is *reflexive* if for all elements $a \epsilon A$, $(a,a) \epsilon R$. $R$ is *symmetric* if for any pair of elements $a,b \epsilon A$, if

$(a,b) \in R$ then $(b,a) \in R$. If $R$ is not symmetric it is called *asymmetric*. $R$ is *antisymmetric* if for any pair of elements $a, b \in A$ if $(a,b) \in R$ then $(b,a) \notin R$. Finally, $R$ is *transitive* if whenever $(a,b) \in R$ and $(b,c) \in R$ then $(a,c) \in R$.

For a relation $R$ on a set $A$, $[a]$ denotes the set of elements to which $a$ is in relation $R$ (i.e., $[a] = \{x \mid aRx\}$. A relation $R$ which is reflexive, symmetric, and transitive is called an *equivalence relation*. We shall now prove that an equivalence relation partitions the set of elements of $A$ into disjoint subsets (i.e., each element of $A$ is in exactly one such subset).

**Theorem 1.1**

If $R$ is an equivalence relation on a set $A$ and $a, b \in A$, then $[a] = [b]$ or $[a] \cap [b] = \emptyset$.

**Proof**

Suppose $[a] \neq [b]$ and $[a] \cap [b] \neq \emptyset$. Let $e_1 \in [a] \cap [b]$. Then $(a,e_1) \in R$ and $(b,e_1) \in R$. Since $[a] \neq [b]$ let $e_2 \in [a]$ and $e_2 \notin [b]$. Then $(a,e_2) \in R$ and $(b,e_2) \notin R$. Since $(b,e_1) \in R$ then $(e_1,b) \in R$ since $R$ is symmetric. Thus $(a,e_1) \in R$ and $(e_1,b) \in R$ and therefore $(a,b) \in R$ (by transitivity) and $(b,a) \in R$ (by symmetry). If $(b,a) \in R$ and $(a,e_2) \in R$, then by transitivity $(b,e_2) \in R$, which contradicts our assumption, and proves the theorem.

### 1.2.3 MAPPINGS

If $A$ and $B$ are two nonempty sets, a *mapping* $\alpha$ from $A$ into $B$ associates with each element $a \in A$, a single unique element $\alpha(a) \in B$. Thus a mapping from $A$ into $B$ is a subset $M$ of $A \times B$ in which (1) if $(a,b) \in M$ and $(a,c) \in M$, then $b = c$ and (2) for all $a \in A$, there exists an element $(a, \alpha(a)) \in M$. Thus the set $R$ of Example 1.3 defines a mapping while the set $R'$ does not define a mapping since $(1,2) \in R'$ and $(1,3) \in R'$.

The concept of ordered pairs can be generalized to that of ordered $n$-tuples in which $(a_1, a_2 \ldots a_n) = (b_1, b_2 \ldots b_n)$ if and only if $a_i = b_i$ for all $i$, $1 \leq i \leq n$. An important type of mapping is that in which $A$ defines a set of ordered $n$-tuples. For example, if $A$ consists of all ordered pairs (2-tuples) in which each pair element is 0 or 1 and $B$ is the set $\{0,1\}$, then a general mapping from $A$ to $B$ is defined by $M = ((0,0), b_0), ((0,1), b_1), ((1,0), b_2), ((1,1), b_3)$ where each of

$b_0, b_1, b_2, b_3$ may be 0 or 1. There are thus $2^4$ possible mappings from $A$ to $B$. Such mappings of $n$-tuples into the set $\{0,1\}$ are important in the design of switching circuits.

## 1.3 BOOLEAN ALGEBRA

The basic mathematical tool which is used in the analysis and synthesis of digital switching circuits is *Boolean algebra*. A Boolean algebra is defined in terms of a finite set* $K$, two binary operations, $+$ (sum) and $\cdot$ (product) and a set of basic postulates. The equality sign $(=)$ is used to indicate that the functional expressions on both sides are equivalent and parentheses are used for ordering multioperand operations in terms of sequences of two-operand operations.

The following are a set of basic postulates, which were first proposed by Huntington[4].

### Huntington's Postulates

1) *Closure:* For all $x, y \epsilon K$
   (a) $x + y \epsilon K$
   (b) $x \cdot y \epsilon K$
2) *Existence of Additive and Multiplicative Identity Elements:*
   (a) There exists an element $0 \epsilon K$ such that $x + 0 = x$, for all $x \epsilon K$.
   (b) There exists an element $1 \epsilon K$ such that $x \cdot 1 = x$, for all $x \epsilon K$.
3) *Commutativity:* For all $x, y \epsilon K$,
   (a) $x + y = y + x$
   (b) $x \cdot y = y \cdot x$
4) *Distributivity:* For all $x, y, z \epsilon K$,
   (a) $x + (y \cdot z) = (x + y) \cdot (x + z)$
   (b) $x \cdot (y + z) = (x \cdot y) + (x \cdot z)$
5) *Complementation:* For every $x \epsilon K$, there exists an element $\bar{x} \epsilon K$ (called the *complement* of $x$) such that (a) $x + \bar{x} = 1$ and (b) $x \cdot \bar{x} = 0$.
6) *Two Distinct Elements:* There exist at least two elements $x, y \epsilon K$ such that $x \neq y$.

---

*This algebra is useful for digital switching circuits when $K$ is restricted to contain exactly two elements. However, we will present the more general formulation in terms of the general set $K$.

Note that the basic postulates are grouped in pairs and that one postulate of each pair can be obtained from the other simply by interchanging all +'s and ·'s, and 0's and 1's. This property is called *duality*.

Many important results of Boolean algebra can be obtained directly from the basic set of postulates. The proofs of these results make use of the *principle of substitution*, if two expressions are equivalent, one may be substituted for the other. The *principle of duality* applies to these results also since if two expressions are proven equivalent, using a sequence of postulates, the duals of these expressions can be proven equivalent using the duals of those postulates. Thus we shall prove only part of each of these results and the other part can be proven by duality.

**Lemma 1.1**
   (a) The element 1 is unique;
   (b) The element 0 is unique.

**Proof**
   (a) Let there be two elements $1_a$ and $1_b$ such that $x \cdot 1_a = x$ and $x \cdot 1_b = x$. Letting $x = 1_b$ in the first equation and $x = 1_a$ in the second, we obtain $1_b \cdot 1_a = 1_b$ and $1_a \cdot 1_b = 1_a$. By commutativity, the left sides of the two equations are equal. Therefore $1_a = 1_b$.
   (b) By duality.

**Lemma 1.2 (Idempotence)**
   For all $x \in K$, (a) $x \cdot x = x$, (b) $x + x = x$.

**Proof**

| | | |
|---|---|---|
| (a) | $x \cdot x = (x \cdot x) + 0$ | Postulate 2(a) |
| | $= (x \cdot x) + (x \cdot \bar{x})$ | Postulate 5(b) |
| | $= x \cdot (x + \bar{x})$ | Postulate 4(b) |
| | $= x \cdot 1$ | Postulate 5(a) |
| | $= x$ | Postulate 2(b) |

   (b) By duality.

**Lemma 1.3**
   For every $x \in K$, (a) $x \cdot 0 = 0$, (b) $x + 1 = 1$.

**Proof**

(a)
$$x \cdot 0 = (x \cdot 0) + 0 \qquad \text{Postulate 2(a)}$$
$$= (x \cdot 0) + (x \cdot \bar{x}) \qquad \text{Postulate 5(b)}$$
$$= x \cdot (0 + \bar{x}) \qquad \text{Postulate 4(b)}$$
$$= x \cdot \bar{x} \qquad \text{Postulate 2(a)}$$
$$= 0 \qquad \text{Postulate 5(b)}$$

(b) By duality.

**Lemma 1.4**
(a) $\bar{0} = 1$ and (b) $\bar{1} = 0$.

**Proof**

(a)
$$\bar{0} + 0 = \bar{0} \qquad \text{Postulate 2(a)}$$
$$\bar{0} + 0 = 1 \qquad \text{Postulate 5(a)}$$

Therefore $\qquad \bar{0} = 1.$

(b) By duality.

**Lemma 1.5**
For every $x \epsilon K$, $\bar{x}$ is unique.

**Proof**
Let $x$ have two complements, $\bar{x}_1$ and $\bar{x}_2$.

$$\bar{x}_1 = \bar{x}_1 \cdot 1 \qquad \text{Postulate 2(b)}$$
$$= \bar{x}_1 \cdot (x + \bar{x}_2) \qquad \text{Postulate 5(a)}$$
$$= (\bar{x}_1 \cdot x) + (\bar{x}_1 \cdot \bar{x}_2) \qquad \text{Postulate 4(b)}$$
$$= 0 + \bar{x}_1 \cdot \bar{x}_2 \qquad \text{Postulate 5(b)}$$
$$= (\bar{x}_2 \cdot x) + (\bar{x}_1 \cdot \bar{x}_2) \qquad \text{Postulate 5(b)}$$
$$= \bar{x}_2 \cdot (x + \bar{x}_1) \qquad \text{Postulate 4(b)}$$
$$= \bar{x}_2 \cdot 1 \qquad \text{Postulate 5(a)}$$
$$= \bar{x}_2 \qquad \text{Postulate 2(b)}$$

**Lemma 1.6 (Absorption)**
 For all $x, y \in K$, (a) $x \cdot (x + y) = x$; (b) $x + (x \cdot y) = x$.

**Proof**

(a) $\qquad\qquad x \cdot (x + y) = (x + 0) \cdot (x + y)$ $\qquad\qquad$ Postulate 2(a)

$\qquad\qquad\qquad\qquad = x + (0 \cdot y)$ $\qquad\qquad\qquad$ Postulate 4(a)

$\qquad\qquad\qquad\qquad = x + 0$ $\qquad\qquad\qquad\qquad$ Lemma 1.3(a)

$\qquad\qquad\qquad\qquad = x$ $\qquad\qquad\qquad\qquad\quad$ Postulate 2(a)

(b) By duality.

**Lemma 1.7**
 For all $x, y \in K$, if (a) $x \cdot y = y$ and (b) $x + y = y$, then $x = y$.

**Proof**
 Substituting equation (b) into equation (a), we obtain the expression

$$x \cdot (x + y) = y$$

But $\qquad\qquad\qquad\qquad x \cdot (x + y) = x$ $\qquad\qquad\qquad$ Lemma 1.6(a)

Therefore $\qquad\qquad\qquad\qquad\qquad\quad x = y.$

**Lemma 1.8**
 For every $x \in K, \bar{\bar{x}} = x$.

**Proof**

$$\bar{\bar{x}} = \bar{\bar{x}} \cdot (\bar{x} + x) \qquad\qquad\qquad \text{Postulate 5(a)}$$

$$= (\bar{\bar{x}} \cdot \bar{x}) + (\bar{\bar{x}} \cdot x) \qquad\qquad \text{Postulate 4(b)}$$

$$= \bar{\bar{x}} \cdot x \qquad\qquad \text{Postulates 5(b) and 2(a)}$$

Therefore $\qquad\quad \bar{\bar{x}} \cdot x = \bar{\bar{x}}.$

$$\bar{\bar{x}} = \bar{\bar{x}} + (x \cdot \bar{x}) \qquad\qquad \text{Postulates 2(a) and 5(b)}$$

$$= (\bar{\bar{x}} + x) \cdot (\bar{\bar{x}} + \bar{x}) \qquad\qquad \text{Postulate 4(a)}$$

$$= (\bar{\bar{x}} + x) \cdot 1 \qquad\qquad\qquad \text{Postulate 5(a)}$$

Therefore $\qquad \bar{\bar{x}} + x = \bar{\bar{x}}.$

Since $\qquad \bar{\bar{x}} \cdot x = \bar{x}$ and $\bar{\bar{x}} + x = \bar{x}$, then

$$\bar{\bar{x}} = x. \qquad\qquad \text{Lemma 1.7}$$

### Theorem 1.2 (Associativity)

For all $x, y, z \epsilon K$, (a) $x \cdot (y \cdot z) = (x \cdot y) \cdot z$;

(b) $x + (y + z) = (x + y) + z.$

### Proof

(a) Let $W_1 = (x \cdot y) \cdot z$, $W_2 = x \cdot (y \cdot z)$, and $W = W_1 + W_2$. We shall prove that $W = W_1$ and $W = W_2$ and therefore $W_1 = W_2$.

$$W = W_1 + W_2$$
$$= (W_1 + x) \cdot (W_1 + (y \cdot z)) \qquad \text{Postulate 4(a)}$$

But $\qquad (W_1 + x) = ((x \cdot y) \cdot z) + x$

$$= [(x \cdot y) + x] \cdot (z + x) \qquad \text{Postulate 4(a)}$$
$$= x \cdot (z + x) \qquad\qquad \text{Lemma 1.6(b)}$$
$$= x \qquad\qquad\qquad \text{Lemma 1.6(a)}$$

Therefore $\qquad W = x \cdot (W_1 + (y \cdot z))$

$$= x \cdot ((W_1 + y) \cdot (W_1 + z)) \qquad \text{Postulate 4(a)}$$
$$= x \cdot (y \cdot z) = W_2 \qquad\qquad \text{Lemma 1.6}$$

Similarly $\qquad W = W_1 + W_2$

$$= [(x \cdot y) + W_2] \cdot (z + W_2) \qquad \text{Postulate 4(a)}$$
$$= [(x + W_2) \cdot (y + W_2)] \cdot (z + W_2) \qquad \text{Postulate 4(a)}$$
$$= (x \cdot y) \cdot z = W_1 \qquad\qquad \text{(Exercise).}$$

Therefore $W_1 = W_2$ which proves the theorem.

(b) By duality.

Because of the associative property, the parentheses grouping consecutive ' $\cdot$ ' or ' $+$ ' operations may be omitted without ambiguity. For convenience of notation, we shall also omit the ' $\cdot$ ' and juxtapose the operands wherever this will not lead to any ambiguity.

**Theorem 1.3 (DeMorgan's law)**
For all $x$, $y \in K$, (a) $\overline{x + y} = \bar{x} \cdot \bar{y}$; (b) $\overline{x \cdot y} = \bar{x} + \bar{y}$.

**Proof**

| | |
|---|---|
| (a)     $(x + y) \cdot (\bar{x} \cdot \bar{y}) = (x \cdot (\bar{x} \cdot \bar{y})) + y(\bar{x} \cdot \bar{y}))$ | Postulate 4(b) |
| $= ((x \cdot \bar{x}) \cdot \bar{y}) + (\bar{x} \cdot (y \cdot \bar{y}))$ | Theorem 1.2(a) |
| $= (0 \cdot \bar{y}) + (\bar{x} \cdot 0) = 0$ | Postulate 5(b) |
| $x + y + \bar{x}\bar{y} = (x(y + \bar{y})) + (y(x + \bar{x})) + \bar{x}\bar{y}$ | Postulate 5(a) |
| $= xy + x\bar{y} + xy + \bar{x}y + \bar{x}\bar{y}$ | Postulate 4(b) |
| $= (xy + xy) + x\bar{y} + \bar{x}y + \bar{x}\bar{y}$ | Theorem 1.2(a) and Postulate 3(a) |
| $= xy + x\bar{y} + \bar{x}y + \bar{x}\bar{y}$ | Lemma 1.2(b) |
| $= (x(y + \bar{y})) + (\bar{x}(y + \bar{y}))$ | Postulate 4(b) |
| $= x \cdot 1 + \bar{x} \cdot 1$ | Postulate 5(a) |
| $= x + \bar{x}$ | Postulate 2(b) |
| $= 1.$ | Postulate 5(a) |

Since $(x + y) \cdot (\bar{x} \cdot \bar{y}) = 0$ and $(x + y) + (\bar{x} \cdot \bar{y}) = 1$, it follows from Postulate 5 that $\overline{x + y} = \bar{x} \cdot \bar{y}$.

(b) By duality.

Since all the above theorems and lemmas relate to elements of a finite set $K$, it is possible to prove them *for any particular set $K$* by enumerating all possible values of each variable. For example, if $K = \{0,1\}$, Theorem 1.3(b) can be proved by using the table of Fig. 1.11,

| $x$ | $y$ | $x \cdot y$ | $\overline{x \cdot y}$ | $\bar{x}$ | $\bar{y}$ | $\bar{x} + \bar{y}$ |
|---|---|---|---|---|---|---|
| 0 | 0 | 0 | 1 | 1 | 1 | 1 |
| 0 | 1 | 0 | 1 | 1 | 0 | 1 |
| 1 | 0 | 0 | 1 | 0 | 1 | 1 |
| 1 | 1 | 1 | 0 | 0 | 0 | 0 |

Figure 1.11   Tabular proof of theorem 1.2

which lists all possible combinations of $x$ and $y$ and also $\overline{x \cdot y}$ and $\bar{x} + \bar{y}$ for each combination. The validity of Theorem 1.3(b) for $K = \{0,1\}$ is established because the columns corresponding to $\overline{x \cdot y}$ and $\bar{x} + \bar{y}$ are identical for all possible values of $x$ and $y$.

Some of the basic laws and postulates of Boolean algebra are summarized in Figure 1.12.

(1) $x + 0 = x$         Identity Elements

      $x \cdot 1 = x$

(2) $x + y = y + x$        Commutativity

      $x \cdot y = y \cdot x$

(3) $x + (y \cdot z) = (x + y)(x + z)$     Distributivity

      $x \cdot (y + z) = x \cdot y + x \cdot z$

(4) $x + \bar{x} = 1$         Complementation

      $x \cdot \bar{x} = 0$

(5) $x \cdot x = x$         Idempotence

      $x + x = x$

(6) $x \cdot 0 = 0$         Special Properties

      $x + 1 = 1$         of 0 and 1

(7) $x \cdot (x + y) = x$       Absorption

      $x + (x \cdot y) = x$

(8) $x + (y + z) = (x + y) + z$     Associativity

      $x \cdot (y \cdot z) = (x \cdot y) \cdot z$

(9) $\overline{x + y} = \bar{x} \cdot \bar{y}$       DeMorgan's Law

      $\overline{x \cdot y} = \bar{x} + \bar{y}$

**Figure 1.12   Basic laws of Boolean algebra**

## Boolean Expressions and Functions

A *Boolean expression* can be recursively defined as follows

(1) $0, 1, x_i, \bar{x}_i$ are Boolean expressions where $x_i$ is any Boolean variable

(2) If $E_1$ and $E_2$ are Boolean expressions then so are $\bar{E}_1, \bar{E}_2, E_1 + E_2, E_1 \cdot E_2$

(3) The only Boolean expressions are those defined by (1) and (2).

The basic laws of Boolean algebra can be used to prove the equivalence of Boolean expressions as illustrated in the following example.

### Example 1.4

We shall prove that the Boolean expressions (1) $(\bar{x}_1 \bar{x}_2 (x_3 x_1 + \bar{x}_2))$ $+ (x_1 + x_2)(x_1 \bar{x}_2 \bar{x}_3 + \bar{x}_1 x_2 x_3)$ and (2) $\bar{x}_2 \bar{x}_3 + \bar{x}_1 x_3$ are equivalent.

Consider the first term of (1), $\bar{x}_1 \bar{x}_2 (x_3 x_1 + \bar{x}_2)$.

This term can be simplified as follows:

$$\bar{x}_1 \bar{x}_2 (x_3 x_1 + \bar{x}_2) = \bar{x}_1 \bar{x}_2 x_3 x_1 + \bar{x}_1 \bar{x}_2 \bar{x}_2 \qquad \text{Distributivity}$$

$$= (\bar{x}_1 x_1) \bar{x}_2 x_3 + \bar{x}_1 \cdot (\bar{x}_2 \bar{x}_2) \qquad \begin{array}{l}\text{Commutativity} \\ \text{and Associativity}\end{array}$$

$$= 0 + \bar{x}_1 \bar{x}_2$$

$$= \bar{x}_1 \bar{x}_2 \; (3)$$

Now consider the second term of (1)

$$(x_1 + x_2)(x_1 \bar{x}_2 \bar{x}_3 + \bar{x}_1 x_2 x_3) = x_1 \cdot (x_1 \bar{x}_2 \bar{x}_3 + \bar{x}_1 x_2 x_3)$$

$$+ x_2 (x_1 \bar{x}_2 \bar{x}_3 + \bar{x}_1 x_2 x_3)$$

$$= x_1 x_1 \bar{x}_2 \bar{x}_3 + x_1 \bar{x}_1 x_2 x_3 + x_2 x_1 \bar{x}_2 \bar{x}_3$$

$$+ x_2 \bar{x}_1 x_2 x_3$$

$$= x_1 \bar{x}_2 \bar{x}_3 + 0 + 0 + \bar{x}_1 x_2 x_3$$

$$= x_1 \bar{x}_2 \bar{x}_3 + \bar{x}_1 x_2 x_3$$

Adding (3) we obtain

$$(1) \quad = \bar{x}_1 \bar{x}_2 + x_1 \bar{x}_2 \bar{x}_3 + \bar{x}_1 x_2 x_3$$

$$= \bar{x}_1 \bar{x}_2 (x_3 + \bar{x}_3) + x_1 \bar{x}_2 \bar{x}_3 + \bar{x}_1 x_2 x_3$$

$$= \bar{x}_1 \bar{x}_2 x_3 + \bar{x}_1 \bar{x}_2 \bar{x}_3 + x_1 \bar{x}_2 \bar{x}_3 + \bar{x}_1 x_2 x_3$$

$$= \bar{x}_1 \bar{x}_2 x_3 + \bar{x}_1 x_2 x_3 + \bar{x}_1 \bar{x}_2 \bar{x}_3 + x_1 \bar{x}_2 \bar{x}_3 \qquad \text{Commutativity}$$

$$= \bar{x}_1 x_3 (\bar{x}_2 + x_2) + \bar{x}_2 \bar{x}_3 (\bar{x}_1 + x_1)$$

$$= \bar{x}_1 x_3 + \bar{x}_2 \bar{x}_3$$

$$= \bar{x}_2 \bar{x}_3 + \bar{x}_1 x_3 = (2)$$

thus proving the equivalence of (1) and (2)

If $x_1, x_2, \ldots, x_n$ is a set of Boolean variables, each of which can assume the values 0 or 1, the $n$-tuple $(x_1, x_2, \ldots, x_n)$ can assume any of $2^n$ possible values. A *completely specified Boolean function* $f(x_1, x_2, \ldots, x_n)$ is a mapping from the set of $2^n$ possible values of this $n$-tuple into $K = \{0,1\}$. Since each of the $2^n$ $n$-tuples can be mapped into either of two possible values there are $2^{2^n}$ possible completely specified Boolean functions.

An *incompletely specified* Boolean function $f'$ $(x_1, x_2, \ldots x_n)$ is a mapping from a subset of the $2^n$ $n$-tuples defined by $(x_1, x_2, \ldots x_n)$ into $K = \{0,1\}$. The value of $f'$ for the remaining $n$-tuples is left unspecified. Since each of the $2^n$ $n$-tuples can be mapped into 0 or 1 or left unspecified there are $3^{2^n}$ possible incompletely specified Boolean functions.

Since a Boolean expression defines a mapping from $(x_1, x_2, \ldots, x_n)$ into $K = \{0,1\}$, every Boolean expression corresponds to a unique (completely specified) Boolean function. De'Morgans laws can be generalized to form the complement of any Boolean expression by *interchanging all $+$'s and $\cdot$'s and complementing all variables*. In this way an expression for the complement of a function $f(x_1, x_2, \ldots, x_n)$ can be obtained from an expression for $f$. Thus if

$$f = (x_1 + (x_2 \cdot x_3)) \cdot (x_4 \cdot (\bar{x}_3 + \bar{x}_5)), \text{ then}$$

$$\bar{f} = (\bar{x}_1 \cdot (\bar{x}_2 + \bar{x}_3)) + (\bar{x}_4 + (x_3 \cdot x_5)).$$

The dual $f_d$, of a combinational function $f(x_1, x_2, \ldots, x_n)$ is defined as $\bar{f}(\bar{x}_1, \bar{x}_2, \ldots, \bar{x}_n)$. It follows from the previous result that an expression for $f_d$ can be obtained from an expression for $f$ simply by *interchanging all $+$'s and $\cdot$'s.* (Problem 1.13($b$)). If $f = (x_1 \cdot (x_2 + (\bar{x}_3 \cdot \bar{x}_4))) + (\bar{x}_1 \bar{x}_2)$ then

$$f_d = (x_1 + (x_2 \cdot (\bar{x}_3 + \bar{x}_4))) \cdot (\bar{x}_1 + \bar{x}_2)*$$

*Many parentheses can be deleted from Boolean expressions under the convention that $\cdot$ dominates $+$. For example, with this convention $x_1 + (x_2 \cdot x_3)$ can be expressed as $x_1 + x_2 x_3$.

A combinational switching function of $n$ variables, $f(x_1,x_2, ...x_n)$, can be expressed as a sum of two functions of $(n - 1)$ variables

$$f = x_i \cdot g_1(x_1, x_2, ...,x_{i-1}, x_{i+1}, ...,x_n)$$
$$+ \bar{x}_i \cdot g_0(x_1,x_2, ...,x_{i-1},x_{i+1}, ...,x_n)$$

where $g_1(x_1,x_2, ...,x_{i-1},x_{i+1}, ...,x_n) = f(x_1,x_2, ...,x_{i-1},1,x_{i+1}, ...,x_n)$

and $g_0(x_1,x_2, ...,x_{i-1},x_{i+1}, ...,x_n) = f(x_1,x_2, ...,x_{i-1},0,x_{i+1}, ...,x_n)$.

This result is known as *Shannon's expansion theorem*[6] (Problem 1.7). The function $f$ can be expanded about any of its variables. This result can be iterated to obtain an expression for $f$ as a sum of $2^k$ terms, each such term being a product of $k$ literals* and a function of $(n - k)$ variables. If $k = n$ each function of $(n - k)$ variables is 0 or 1, and the function $f$ is expressed as a sum of $k$-literal product terms. This expression, called the *sum of minterms* expression for $f$, is of importance in combinational circuit design as we shall see in Chapter 3. Shannon's expansion theorem is also useful for proving results about properties of functions which we shall consider in Chapter 4.


## 1.4 APPLICATION OF BOOLEAN ALGEBRA TO SWITCHING CIRCUITS

The output of a switching circuit with inputs $(x_1,x_2... x_n)$ and output $z$ is called a switching function. If the output only depends on the present value of the inputs and is completely independent of previous (past) values of the inputs, the circuit is said to be a *combinational circuit* and the switching function a *combinational function*. If the output depends on previous values of the inputs as well as the present values the circuit is a *sequential circuit* and the function is a *sequential function*.

The two-valued Boolean algebra, $K = \{0,1\}$, is useful in describing digital circuits. The signal on any lead in the circuit can be represented by a Boolean variable which may assume the value of 0 or 1. The behavior of the different types of gates shown in Figures 1.7 and 1.8 can be described by Boolean expressions. Since digital circuits are formed by interconnecting such gates, Boolean algebra may be used in analyzing their behavior.

---

*A literal is a variable $x_i$ or the complement of a variable $\bar{x}_i$.

If $K = \{0,1\}$ the Boolean product as defined by postulate 2(b) and Lemma 1.3(a) yields $0 \cdot 0 = 0 \cdot 1 = 1 \cdot 0 = 0$ and $1 \cdot 1 = 1$ which is identical to the logical AND function defined by Figure 1.7. Thus if we use the Boolean variables $x_1$ and $x_2$ to represent the inputs to the two-input AND gate of Figure 1.7 and $z$ to represent its output, then the output of the gate may be written as $z = x_1 \cdot x_2$. Similarly the Boolean sum is defined by $0 + 0 = 0$, $0 + 1 = 1 + 0 = 1 + 1 = 1$, which corresponds to the logical OR function. Thus the output of a 2-input OR gate whose inputs are represented by Boolean variables $x_1$ and $x_2$, may be represented as $x_1 + x_2$. Similarly, the output of a NOT gate with input $x$ is $z = \bar{x}$. The NAND and NOR gate outputs are represented by $z = \overline{x_1 x_2} = \bar{x}_1 + \bar{x}_2$ and $z = \overline{x_1 + x_2} = \bar{x}_1 \cdot \bar{x}_2$ respectively.

The output of any two-input gate is not affected if the inputs $x_1$ and $x_2$ are interchanged. Because of the commutative property of $\cdot$ and $+$, the values of the Boolean expressions representing these gates are not changed by changing the order of the input variables.

Similarly, the output of multiple input gates is not affected by the ordering of its inputs. Because of the commutative and associative properties of the Boolean algebra, the expressions representing the output of such gates are not changed by reordering or regrouping the input variables. Thus parentheses can be omitted from these expressions. The outputs of 3-input gates with input variables $x_1, x_2, x_3$ can be represented by $x_1 x_2 x_3$, $x_1 + x_2 + x_3$, $\overline{x_1 x_2 x_3} = \bar{x}_1 + \bar{x}_2 + \bar{x}_3$, and $\overline{x_1 + x_2 + x_3} = \bar{x}_1 \bar{x}_2 \bar{x}_3$ for AND, OR, NAND and NOR gates respectively.

## 1.5 ANALYSIS OF COMBINATIONAL CIRCUITS

The *analysis of combinational circuits* consists of determining the function realized by a circuit. The description of the function may be in the form of a Boolean expression or in the form of a table specifying the output of the function for every possible input combination. The latter is called the *truth table** of the function. Any combinational function

---

*The term truth table has its origins in mathematical logic where a similar table is sometimes used to determine whether a given logical expression $E$ is *true or false* depending upon whether the variables of $E$ are true or false.

can be realized by a circuit without feedback loops** (called an *acyclic circuit*). Such circuits can be analyzed very easily using the basic laws of 2-valued Boolean algebra, presented in the previous section, for expression simplification.

Two procedures for analyzing an acyclic circuit will be demonstrated for the circuit of Figure 1.13. The output function of gates all of whose inputs are input variables $(x_i)$ may be determined by inspection. Denoting by $G_i$, the output of the gate with that label, we have $G_1 = x_1 + x_2$ and $G_2 = \overline{x_3 + x_4} = \bar{x}_3 \bar{x}_4$. After this the outputs of all gates whose inputs are input variables or gates whose outputs were previously

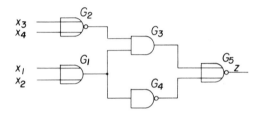

**Figure 1.13    A combinational logic circuit**

determined can be represented and simplified expressions can be obtained using elementary rules of Boolean algebra. This is repeated until an expression for the circuit output is generated. For the circuit of Figure 1.13

$$G_3 = G_2 \cdot G_1 = \bar{x}_3 \bar{x}_4 (x_1 + x_2) = \bar{x}_3 \bar{x}_4 x_1 + \bar{x}_3 \bar{x}_4 x_2$$
$$G_4 = \bar{G}_1 = \overline{(x_1 + x_2)} = \bar{x}_1 \bar{x}_2$$
$$z = G_5 = \overline{G_3 + G_4} = \bar{G}_3 \bar{G}_4 = \overline{\bar{x}_3 \bar{x}_4 x_1 + \bar{x}_3 \bar{x}_4 x_2} \cdot \overline{\bar{x}_1 \bar{x}_2}$$
$$= (x_3 + x_4 + \bar{x}_1) \cdot (x_3 + x_4 + \bar{x}_2) \cdot (x_1 + x_2)$$
$$= x_1 x_3 + x_1 x_4 + x_2 x_3 + x_2 x_4$$

The output has been expressed as a sum of product terms and the

*A circuit $C$ has a feedback loop if there exists an ordered set of gates $\{G_1, G_2 ... G_r\}$ in $C$ such that the output of $G_i$ is an input to $G_{i+1}$ for all $i$, $1 \le i \le r - 1$, and the output of $G_r$ is an input to $G_1$. Informally, there is a directed path from the output of a gate $G_1$ in $C$ to an input to $G_1$.

output of the circuit is 1 if and only if all the input variables in some product term are 1. Thus, the output of the circuit is 1 if $x_3 = 1$ and $x_1$ or $x_2 = 1$ or if $x_4 = 1$ and $x_1$ or $x_2 = 1$.

The circuit can also be analyzed by working backwards from the outputs to the inputs. For the circuit of Figure 1.13 we obtain the following sequence of equations:

$$z = G_5 = \overline{G_4 + G_3} = \bar{G}_4 \bar{G}_3 \tag{1}$$

$$G_4 = \bar{G}_1 \tag{2}$$

$$G_3 = G_1 \cdot G_2 \tag{3}$$

Substituting equations (2) and (3) in equation (1) we obtain:

$$z = \bar{\bar{G}}_1 \overline{(G_1 \cdot G_2)} = G_1 (\bar{G}_1 + \bar{G}_2) = G_1 \bar{G}_2 \tag{1'}$$

Now continuing we obtain:

$$G_1 = x_1 + x_2 \tag{4}$$

$$G_2 = \overline{x_3 + x_4} = \bar{x}_3 \bar{x}_4 \tag{5}$$

Substituting (4) and (5) in (1') we obtain:

$$z = (x_1 + x_2)(\overline{\bar{x}_3 \bar{x}_4}) = (x_1 + x_2)(x_3 + x_4)$$
$$= x_1 x_3 + x_1 x_4 + x_2 x_3 + x_2 x_4$$

which is the same result as we obtained previously.

The value of the output for any input combination can be determined by substituting the input values in the Boolean expression for $z$. The truth table of Figure 1.14 can be obtained from the final expression in this manner.

The truth table has a row for every possible input combination. Thus the truth table of an $n$-variable function will have $2^n$ rows, since each variable can assume either of two values.

In Chapter 3 we shall consider the problem of synthesizing a combinational circuit to realize a Boolean function specified in the form of a truth table.

| $x_1$ | $x_2$ | $x_3$ | $x_4$ | $z$ |
|---|---|---|---|---|
| 0 | 0 | 0 | 0 | 0 |
| 0 | 0 | 0 | 1 | 0 |
| 0 | 0 | 1 | 0 | 0 |
| 0 | 0 | 1 | 1 | 0 |
| 0 | 1 | 0 | 0 | 0 |
| 0 | 1 | 0 | 1 | 1 |
| 0 | 1 | 1 | 0 | 1 |
| 0 | 1 | 1 | 1 | 1 |
| 1 | 0 | 0 | 0 | 0 |
| 1 | 0 | 0 | 1 | 1 |
| 1 | 0 | 1 | 0 | 1 |
| 1 | 0 | 1 | 1 | 1 |
| 1 | 1 | 0 | 0 | 0 |
| 1 | 1 | 0 | 1 | 1 |
| 1 | 1 | 1 | 0 | 1 |
| 1 | 1 | 1 | 1 | 1 |

Figure 1.14    Truth table for circuit of Figure 1.13

## SOURCES

The basic digital circuit realizations of gates can be found in Millman and Taub[5]. There are many books which cover the material on set theory. Among these are Birkhoff and MacLane[1]. The set of postulates for Boolean algebra were formulated by Huntington[4]. and simplified proofs of some of the basic results were presented in Hill and Peterson[3]. Shannon[6] developed the algebra of switching circuits and demonstrated its relation to Boolean algebra[2].

## REFERENCES

[1] Birkhoff, G., and S. MacLane, *A Survey of Modern Algebra,* 3rd edition, The Macmillan Company, New York, N.Y., 1965.

30    Fundamental Concepts

[2]  Boole, G., *An Investigation of the Laws of Thought*, Dover Publications, Inc., New York, N.Y., 1854.

[3]  Hill, F. J., and G. R. Peterson, *Introduction to Switching Theory and Logical Design*, J. Wiley & Sons, New York, N.Y., 1968.

[4]  Huntington, E. V., "Sets of Independent Postulates for the Algebra of Logic," *Trans. American Mathematical Society*, vol. 5, pp. 288-309, 1904.

[5]  Millman, J. and H. Taub, *Pulse, Digital and Switching Waveforms*, McGraw-Hill, New York, N.Y., 1965.

[6]  Shannon, C. E., "A Symbolic Analysis of Relay and Switching Circuits," *Transactions AIEE*, vol. 57, pp. 713-723, 1938.

## PROBLEMS

**1.1)** For a circuit consisting of binary devices where the two voltage (current) levels are $V_H$, $V_L$, $V_H > V_L$, if $V_H$ is represented as 1 and $V_L$ as 0 the circuit is said to have *positive logic* and if $V_H$ is 0, $V_L$ is 1, the circuit is said to have *negative logic*. Consider a device which operates as specified by the following table.

| Inputs | | Output |
|---|---|---|
| $x_1$ | $x_2$ | $f(x_1,x_2)$ |
| $V_L$ | $V_L$ | $V_H$ |
| $V_L$ | $V_H$ | $V_H$ |
| $V_H$ | $V_L$ | $V_H$ |
| $V_H$ | $V_H$ | $V_L$ |

a)  For positive logic what type of gate is this device?
b)  For negative logic what type of gate is this device?
c)  Define a 2-input 1-output device which realizes the same element for positive and negative logic.

**1.2)** Use the basic laws of Boolean algebra to prove (or disprove) the following equivalences.

a) $xy + \bar{x}\bar{y} + \bar{x}yz = xy\bar{z} + \bar{x}\bar{y} + yz$

b) $xyz + w\bar{y}\bar{z} + wxz = w\bar{y}\bar{z} + wx\bar{y} + xyz$

**1.3)** A Boolean function $f(\mathbf{x})$ is self dual if $f = f_d = \overline{f(\bar{\mathbf{x}})}$, where $\mathbf{x}$ denotes a set of variables $(x_1, x_2, \ldots, x_n)$.

Determine if the following functions are self dual.

a) $x_1 \bar{x}_2 + \bar{x}_1 x_2 = f_1(x_1, x_2)$

b) $\bar{x}_2(x_1 + \bar{x}_3) + x_1(\bar{x}_2 + \bar{x}_3) = f_2(x_1, x_2, x_3)$

**1.4)** For each of the following expressions determine the complement expression.

a) $(x_1 + x_2)(\bar{x}_1 x_3 + x_4(x_2 + x_3)(x_3 + x_5)) + \bar{x}_1 \bar{x}_2 \bar{x}_3(x_4 + x_5 x_6)$

b) $\bar{x}_2(x_1 + \bar{x}_3 x_4(\bar{x}_1 + x_2))$

**1.5)** Derive Boolean expressions for the outputs of the circuits of Figure, 1.15(a) and (b). Are these two output expressions equivalent?

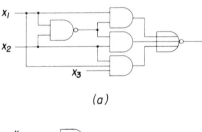

(a)

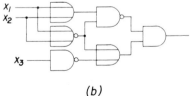

(b)

**Figure 1.15    Problem 1.5**

**1.6)** Define a Boolean algebra with more than two elements. Specify the set of elements, $K$, and define $x \cdot y$, $x + y$, $\bar{x}$, for all $x$, $y \epsilon K$.

**1.7)** Prove that any combinational function can be represented as

a) $f(x_1, x_2, ..., x_n) = x_1 f(1, x_2, ..., x_n) + \bar{x}_1 f(0, x_2, ..., x_n)$

b) $f(x_1, x_2, ..., x_n) = x_1 x_2 f(1, 1, x_3, ..., x_n) + \bar{x}_1 x_2 f(0, 1, x_3, ..., x_n)$
$+ x_1 \bar{x}_2 f(1, 0, x_3, ..., x_n) + \bar{x}_1 \bar{x}_2 f(0, 0, x_3, ..., x_n)$

c) Generalize this expansion to $i$ variables. (This result is known as the *Shannon expansion theorem*).

**1.8)** Use Venn diagrams to prove the following

a) $A \cap (B \cup C) = (A \cap B) \cup (A \cap C)$

b) $(A \cup B) \cap (\bar{A} \cup \bar{B}) = (A \cap \bar{B}) \cup (\bar{A} \cap B)$

c) $I - (A \cup B) = (I - A) \cap (I - B) = \bar{A} \cap \bar{B}$

**1.9)** If $A$ is a set, the *power set of A*, $P(A)$, is the set of all subsets of $A$. If $A$ is a set with $n$ elements, show that $P(A)$ has $2^n$ elements.

**1.10)** For each subset of the properties reflexive, symmetric, transitive, define a relation having those properties but not having any of the other properties.

**1.11)** Consider a symmetric and transitive relation $R$ defined on a set $A$. Prove that if for any $a \in A$, there exists an element $b \in A$ such that $aRb$, then $R$ is reflexive.

**1.12)** Show that the algebra of sets satisfies the fundamental postulates of a Boolean algebra where the $\cup$ set operation corresponds to $+$ and $\cap$ corresponds to $\cdot$ .

**1.13)** (a) From DeMorgan's laws prove that the complement of a Boolean expression can be obtained by interchanging all $+$'s and $\cdot$'s and complementing all variables

(b) Use (a) to prove that the dual of a Boolean expression is obtained by interchanging all $+$'s and $\cdot$'s

# CHAPTER 2

# Number Systems, Nondecimal Arithmetic, and Codes

In human communication, numbers are represented using a decimal number system. In digital systems, signals are usually restricted to two possible values and numbers are frequently represented in a binary (base 2) number system. Since such systems are designed by people, it is important to be able to readily convert information from one number base to another.

## 2.1 NUMBER SYSTEMS

For any positive integer $R$ we can represent numbers in a base $R$ number system. Each such number consists of a sequence of integers $r_n r_{n-1} r_{n-2} \ldots r_1 r_o$, $0 \le r_i \le R - 1$, which represents the decimal number equal to $\sum_{i=0}^{n} r_i \cdot R^i$. Thus in base 6, 543 represents the number equal to $5 \times 6^2 + 4 \times 6^1 + 3 \times 6^0 = 180 + 24 + 3 = 207$. We shall denote this correspondence as $543_6 = 207_{10}$, where subscripts are used to denote the relevant number base. If there is no subscript for a number, it will be assumed to be base 10. In this way it is easy to convert an integer from an arbitrary number base to decimal (base 10).

It is also possible to represent fractional parts of numbers in arbitrary base $R$ number systems. The sequence of integers $r_{-1} r_{-2} \ldots r_{-k}$, preceded by a point (frequently referred to as a decimal point in base 10 number system) represents the number equal to $\sum_{i=-1}^{-k} r_i 2^i$ where $0 \le r_i \le R - 1$. Thus $.431_5$ is equal to $4 \times 5^{-1} + 3 \times 5^{-2} + 1 \times 5^{-3} = .800 + .120 + .008 = .9280_{10}$.

When representing a number in a base $R > 10$, it is necessary to use parentheses around any number $r_i$ in the sequence if $r_i \geq 10$, in order to distinguish $r_i$ from a sequence of two or more numbers. Thus, $111_{12} = 1 \times 12^2 + 1 \times 12^1 + 1 \times 12^0 = 157$, while $(11)1_{12} = (11) \times 12^1 + 1 \times 12^0 = 133$.

Given an integer $N$, we may wish to express $N$ as a base $R$ number. That is, we wish to convert $N$ from base 10 to base $R$, by determining the coefficients $b_k b_{k-1} \ldots b_1 b_0$ where $0 \leq b_i \leq R - 1$ and $\Sigma_{i=0}^{k} b_i \cdot R^i = N$.

$$\sum_{i=0}^{k} b_i \cdot R^i = \left( R \cdot \sum_{i=1}^{k} b_i \cdot R^{i-1} \right) + b_0 = N$$

Dividing by $R$ yields

$$\sum_{i=1}^{k} b_i \cdot R^{i-1} + \frac{b_0}{R} = \frac{N}{R}$$

Hence $b_0$ is the remainder resulting from dividing $N$ by $R$. The quotient of this division will be denoted as $Q_1$ where $Q_1 = \Sigma_{i=1}^{k} b_i \cdot R^{i-1}$. Dividing $Q_1$ by $R$ results in a remainder equal to $b_1$ and a quotient, $Q_2$. This procedure is iterated, the division of $Q_i$ producing a remainder equal to $b_i$, and a quotient $Q_{i+1}$, until $Q_{i+1} = 0$. Thus the following procedure can be used to convert any integer $N$ from base 10 to base $R$.

## Procedure 2.1
(1) Divide $N$ by $R$ resulting in a quotient $Q_1$ and a remainder $P_0 = b_0$, $0 \leq b_0 \leq R - 1$.

(2) Repeat (1) on $Q_i$ resulting in a quotient $Q_{i+1}$ and a remainder $P_i = b_i$, for all $i \geq 1$; terminate when $Q_i = 0$.

## Example 2.1
(a) Convert $152_{10}$ to base 3.
  Dividing 152 by 3 yields $Q_1 = 50$, $P_0 = b_0 = 2$.
  Dividing $Q_1 \models 50$ by 3 yields $Q_2 = 16$, $P_1 = b_1 = 2$.
  Dividing $Q_2 = 16$ by 3 yields $Q_3 = 5$, $P_2 = b_2 = 1$.
  Dividing $Q_3 = 5$ by 3 yields $Q_4 = 1$, $P_3 = b_3 = 2$.
  Dividing $Q_4 = 1$ by 3 yields $Q_5 = 0$, $P_4 = b_4 = 1$.

Therefore, $152_{10} = 12122_3$. This repeated division can be represented in the following format.

Remainders

$$
\begin{array}{r|l}
3 & 152 \\
\hline
& 50 \qquad 2 = b_0 \\
\hline
& 16 \qquad 2 = b_1 \\
\hline
& 5 \qquad\, 1 = b_2 \\
\hline
& 1 \qquad\, 2 = b_3 \\
\hline
& 0 \qquad\, 1 = b_4
\end{array}
$$

(b) Convert $152_{10}$ to base 13.

$$
\begin{array}{r|l}
13 & 152 \\
\hline
& 11 \quad 9 = b_0 \\
\hline
& 0 \;\; 11 = b_1
\end{array}
$$

Therefore $152_{10} = (11)9_{13}$.

A similar procedure may be utilized for converting base 10 fractional numbers to base $R$. Suppose $N = b_{-1} b_{-2}...b_{-p} = \Sigma_{i=-1}^{-p} b_i R^i,\ 0 \le N < 1$.

$$
\sum_{i=-1}^{-p} b_i R^i = R^{-1} \cdot \left[ \sum_{i=-2}^{-p} b_i R^{(i+1)} + b_{-1} \right]
$$

Therefore

$$
N \cdot R = \sum_{i=-2}^{-p} b_i \cdot R^{(i+1)} + b_{-1}
$$

Thus, the multiplication $N \cdot R$ results in a number whose integer part, $I_{-1}$, is equal to $b_{-1}$ and whose fractional part, $F_1$, is $\Sigma_{i=-2}^{-p} b_i R^{(i+1)}$.

Similarly, multiplying $F_1$ by $R$ yields a number whose integer part, $I_2$, is equal to $b_{-2}$ and whose fractional part is $F_2$. This is iterated, the multiplication of $F_k$ by $R$ generating $b_{-(k+1)}$ until $F_k = 0$.

**Procedure 2.2**

(1) To convert a fractional number $.N_{10}$ to base $R$, multiply $.N$ by $R$, resulting in an integer part $I_1 = b_{-1}$ and a fractional part $F_1$.

(2) Repeat (1) on $F_i$ resulting in an integer part $I_{i+1} = b_{-(i+1)}$ and a fractional part $F_{i+1}$, for $i = 1, 2, \ldots$. Terminate when $F_i = 0$.

However, unlike the integer conversion procedure (Proc. 2.1), the fractional number conversion procedure (Proc. 2.2) may not terminate because the number $.N$ may not have a finite representation in base $R$. If $F_j = F_k$ for $j > k$, then the base $R$ representation of $.N$ repeats the bit sequence $b_{-(k+1)}b_{-(k+2)}\ldots b_{-j}$ indefinitely. Note that a nonterminating fractional number in base 10 may be terminating in base R. However, the reverse situation may also occur.

**Example 2.2**

(a) Convert $N = .625_{10}$ to base 2.

Multiplying $N$ by 2 yields 1.25. Thus $F_1 = .25$ and $b_{-1} = I_1 = 1$.

Multiplying $F_1$ by 2 yields .50. Thus, $b_{-2} = 0$ and $F_2 = .50$.

Multiplying $F_2$ by 2 yields 1.00. Thus, $b_{-3} = 1$ and $F_3 = .00$ and the procedure terminates, yielding $.625_{10} = .101_2$.

This repeated multiplication will be represented in the following format.

$$
\begin{array}{rc}
\text{Integer} & .625 \\
\text{Part} & \underline{\quad 2\quad} \\
b_{-1} = 1 & 1\overline{\rvert.25} \\
& \underline{\quad 2\quad} \\
b_{-2} = 0 & 0\overline{\rvert.50} \\
& \underline{\quad 2\quad} \\
b_{-3} = 1 & 1\overline{\rvert.00}
\end{array}
$$

(b) Convert $.N = .6666\ldots_{10}$ to base 3. Multiplying $.N$ by 3 yields

2.00. Thus $b_{-1} = 2$, $F_1 = 0$, and the procedure terminates. Thus $.6666..._{10} = .2_3$.

(c)  Convert .4 to base 2.

```
        .4
        2
   ┌──────────
  0│ .8
        2
   ┌──────────
  1│ .6
        2
   ┌──────────
  1│ .2
        2
   ┌──────────
  0│ .4  ◄── repeat
```

Thus the repeated multiplication yields the result

$$.4_{10} = .0\underbrace{1100110}\ ..._2$$

repeated
sequence

Since the representation of fractional numbers may be nonterminating, it is important to determine the maximum error involved in truncation after $k$ bits. That is, if $N = \Sigma_{i=-1}^{-\infty} b_i R^i$ is represented as $N' = \Sigma_{i=-1}^{-k} b_i R^i$, what is the maximum value of $N - N'$. Since $|N - N'| = \Sigma_{i=-(k+1)}^{-\infty} b_i R^i$, $0 \le b_i \le R - 1$, the maximum value of $N - N'$ occurs if $b_i = R - 1$ for all $i > k$. In this case $|N - N'| < R^{-k}$ (Exercise).

**Example 2.3**

To represent .4 in base 2 using a sufficient number of bits so the error is $< .01$, the number of bits required is the smallest value of $k$ such that $2^{-k} < .01$. This implies $k = 7$. From Example 2.2(c), $.4 = .0110011_2$.

In converting a number with both integer and fractional parts to base $R$, Procedure 2.1 is used to convert the integer part of the number and Procedure 2.2 is used to convert the fractional part of the number. To convert a number from base $R'$ to base $R$, we first convert from base $R'$ to base 10 and then convert from base 10 to base $R'$ using Procedures 2.1 and 2.2.

**Example 2.4**

Convert $26.5_8$ to base 3. Converting to base 10, $26.5_8 = 2 \times 8^1 + 6 \times 8^0 + 5 \times 8^{-1} = 22.625_{10}$. Procedure 2.1 is used to convert the integer 22 to base 3 as follows

Remainders

| 3 | 22 | |
|---|----|---|
| | 7 | $1 = b_0$ |
| | 2 | $1 = b_1$ |
| | 0 | $2 = b_2$ |

Thus, $22 = 211_3$. Procedure 2.2 is used to convert .625 to base 3 as follows

| Integer Parts | .625 | |
|---|---|---|
| | 3 | |
| $b_{-1} = 1$ | 1 \| .875 | repeated |
| | 3 | |
| $b_{-2} = 2$ | 2 \| .625 | |

Thus, $.625 = .1212..._3$ and $26.5_8 = 211.1212..._3$

In digital systems, most signals are restricted to two possible values, so the binary (base 2) number system is important in representing numbers. However, binary representation of numbers has several disadvantages when used by designers who are more familiar with decimal number representation. One of these is the relative difficulty of converting between binary and decimal. Numerical information can also be represented by a sequence of 0,1 integers in such a way that it is simple to convert between the 0,1 representation and decimal. This representation is referred to as *Binary Coded Decimal* (BCD). Each decimal digit is represented by the base 2 number system equivalent sequence of 4 bits. Thus 509 is represented as 0101 0000 1001. To convert from BCD
                                                              5    0    9

to decimal, we simply decode successive 4-bit sequences starting at the left of the word. Thus, 100001110010 is equal to $872_{10}$. Another problem caused by binary number system representation is that for a given number, the length of the sequence is much greater than that required in decimal representation and hence there is much greater chance of error in handling this lengthier sequence. For instance, the 4-digit decimal number 7468 corresponds to the 13-bit binary number system sequence 1110100101100. Numbers represented in the binary number system can be easily transformed to octal (base 8), by translating 3-bit sequences from left to right into octal numbers. Thus, 1(110)(100)(101)(100) = $16454_8$. (In general we can convert from base $R$ to base $R^K$ by translating length $K$ subsequences. Exercise.) This reduces the length of the sequence to one-third that of the binary sequence and correspondingly decreases the probability of error. The reverse transformation is performed by encoding each octal number as a 3-bit binary number. Thus, $7520_8$ = $\underset{7}{111}\ \underset{5}{101}\ \underset{2}{010}\ \underset{0}{000}\ _2$.

It may also be necessary to represent non-numeric information using the two symbols 0,1. Each element which must be represented can be made to correspond to a unique sequence of bits. If the total number of distinct elements is $k$ and all elements correspond to equal length bit sequences, then the length of the sequence $l$ must be such that $2^l \geq k$ or equivalently $l \geq \lceil \log_2 k \rceil$ where $\lceil X \rceil$ is the smallest integer $\geq X$. In decimal number systems each digit can have 10 values. The number of bits required to represent a digit is hence $l \geq \lceil \log_2 10 \rceil$ = 4. Note that six of the 16 4-bit sequences do not correspond to digits. It is possible to use these sequences to represent letters so that alphanumeric information can be represented. The *hexadecimal code* represents the ten numbers, 0-9, and the letters $A,B,C,D,E,F$ as shown in Figure 2.1, and is a generalization and extension of the BCD code previously considered. Using this code, $A15$ is represented as $\underset{A}{1010}\ \underset{1}{0001}\ \underset{5}{0101}$

while $\underset{C}{1100}\ \underset{A}{1010}\ \underset{D}{1101}$ represents *CAD*.

## 2.2 NONDECIMAL ARITHMETIC

In this section we will specify algorithms by which the standard arithmetic operations can be performed in bases other than decimal. The design of binary arithmetic logic circuits is very important in digital system design, since these operations are very common.

| Element | Hexadecimal Representation |
|---------|---------------------------|
| 0 | 0000 |
| 1 | 0001 |
| 2 | 0010 |
| 3 | 0011 |
| 4 | 0100 |
| 5 | 0101 |
| 6 | 0110 |
| 7 | 0111 |
| 8 | 1000 |
| 9 | 1001 |
| A | 1010 |
| B | 1011 |
| C | 1100 |
| D | 1101 |
| E | 1110 |
| F | 1111 |

**Figure 2.1   Hexadecimal Code**

### 2.2.1 BASE $R$ ADDITION

Let us consider the addition of two base $R$ numbers. As in decimal addition, the numbers to be added are aligned accordingly to the (rightmost) least significant integers which are then added. If the two numbers are $A = a_n a_{n-1}...a_1 a_o$ and $B = b_n b_{n-1}...b_1 b_o$ where $0 \le a_i, b_i \le R - 1$ then the result, $A + B = d_{n+1} d_n...d_1 d_o, 0 \le d_i \le R - 1$, is computed as follows. The addition of the least significant positions $a_o, b_o$ results in $a_o + b_o$ where $0 \le a_o + b_o \le 2R - 2$. If $a_o + b_o \ge R$, the sum can be represented in base $R$ by the sequence of two integers $1d_o$ where $d_o = (a_o + b_o) - R$. In this case the addition of the least significant position results in a *carry* of 1 to the next least significant position. If $0 \le a_o + b_o \le R - 1$, the sum is represented in base $R$ as $0d_o$ where $d_o = a_o + b_o$, thus resulting in a 0 carry. If a 1 carry is generated it is then added to the next position addition resulting in $a_1 + b_1 + 1$, which is $\le 2R - 1$. If $a_1 + b_1 + 1 \ge R$, it is represented by $1d_1$

where $d_1 = (a_1 + b_1 + 1) - R$, thus generating a 1 carry into the next position. This is repeated for all positions proceeding from right to left sequentially. Note that the carry into any position is always either 0 or 1. The addition of two numbers represented by integer sequences of length $n$ may result in a number represented by an integer sequence of length $n + 1$.

**Procedure 2.3**

(1) The addition of two base $R$ numbers $A = a_n a_{n-1}...a_o$ and $B = b_n b_{n-1}...b_o$ results in $A + B = d_{n+1} d_n...d_1 d_o$. To compute $d_o$ add $a_o + b_o$. If $a_o + b_o < R$, $d_o = a_o + b_o$, and a carry $C_1 = 0$ is generated for position 1. If $a_o + b_o \geq R$, then $d_o = a_o + b_o - R$ and a 1 carry is generated into position 1.

(2) Add digits* $a_i, b_i$ and the carry $C_i$ into position $i$. If $a_i + b_i + C_i < R$, then $d_i = a_i + b_i + C_i$ and a carry $C_{i+1} = 0$ is generated for the position $i + 1$. If $a_i + b_i + C_i \geq R$, then $d_i = (a_i + b_i + C_i) - R$ and a carry $C_{i+1} = 1$ is generated into position $i + 1$. Repeat sequentially for positions $i = 2,3,...n$.

(3) The most significant digit of $A + B$, $d_{n+1} = C_{n+1}$ where $C_{n+1}$ is the carry resulting from the addition of the $n^{th}$ position numbers $a_n, b_n, C_n$.

**Example 2.5**

(a) Consider the addition of the two base 5 numbers $4323_5$ and $1204_5$. The least significant digits added yield $(3 + 4) = 7$ which is represented in base 5 as 12. Hence there is a carry $C_1 = 1$ and a least significant sum digit $d_o = 2$. The next rightmost digit additions yields $(2 + 0)$ + $\underbrace{1}_{\text{carry}}$ $= 3$, yielding an output $d_1 = 3$, and a carry $C_2 = 0$. The 3rd digit addition is $(3 + 2) + 0 = 5$ which is represented in base 5 as 10, thus generating the 3rd digit output $d_2 = 0$ and the carry $C_3 = 0$. The 4th digit addition is $(4 + 1) + \underbrace{1}_{\text{carry}} = 6$ which generates the output $d_3 = 1$ and the carry $C_4 = 1$. The output $d_4 = C_4 = 1$. The addition can be represented in the following format:

---

*We use the term digits for the individual integers in the base $R$ representation of a number independent of $R$. The term *bit* for binary digit is frequently used for $R = 2$.

$$
\begin{array}{lll}
4323 & 1^{st} & \text{number} \\
1204 & 2^{nd} & \text{number} \\
\underline{1101} & \leftarrow & \text{carries generated} \\
11032 & & \text{sum}
\end{array}
$$

(b) The base 2 addition of 1001 and 0101 is represented as follows:

$$
\begin{array}{lll}
1001 & 1^{st} & \text{number} \\
0101 & 2^{nd} & \text{number} \\
\underline{0001} & & \text{generated carries} \\
01110 & & \text{sum}
\end{array}
$$

In base $R$ addition of numbers with fractional parts, the least significant positions of the integer parts of the operands are aligned and the numbers are added in serial manner, least significant bit first.

## 2.2.2 BASE $R$ SUBTRACTION

Subtraction can be performed in a manner similar to addition, requiring alignment of the least significant positions of the integer parts of the operands, and successive digit subtraction proceeding from least to most significant position. The base $R$ subtraction $A - B$ where $A = a_n \ldots a_1 a_0$, and $B = b_n \ldots b_1 b_0$ results in $D = d_n \ldots d_1 d_0$ where $0 \le a_i, b_i \le R - 1$. Since $-(R - 1) \le a_o - b_o \le (R - 1)$ and $0 \le d_0 \le R - 1$, if $a_o - b_o < 0$, then we borrow (subtract) a 1 from the rightmost nonzero digit $a_k$ in $A$, lend (add) $(R - 1)$ to all intermediate digits between $a_k$ and $a_o$, and add $R$ to bit $a_o$ thus generating $d_o = (a_o + R) - b_o$. (The validity of this borrowing/lending technique stems from the fact that $x \cdot R^k = (x - 1) R^k + (\Sigma_{i=1}^{k-1} (R - 1) R^i) + R$. (Exercise).) If $a_o - b_o \ge 0$ then $d_o = a_o - b_o$. We then repeat this procedure for all digit positions proceeding from right to left sequentially.

## Procedure 2.4

(1) If $A = a_n a_{n-1} \ldots a_1 a_o$ and $B = b_n b_{n-1} \ldots b_1 b_o$ are base $R$ numbers then $A - B = d_n d_{n-1} \ldots d_1 d_o$ is derived by successive digit subtraction as follows. If $a_i - b_i \ge 0$ then $d_i = a_i - b_i$. If $a_i - b_i < 0$ then $d_i = (a_i + R) - b_i$ and a 1 is borrowed (subtracted) from the rightmost nonzero digit $a_k$, $k > i$, and $R - 1$ is added to all digits $a_{k-1}, a_{k-2} \ldots a_{i+1}$.

(2) Step (1) is repeated for $i = 1,2,...,n$ to obtain the base $R$ representation of $A - B$.

**Example 2.6**

Subtract $4625_7$ from $6016_7$. The least significant digit subtraction yields $(6 - 5) = 1 = d_o$ as the first output digit. The next least significant digit subtraction is $(1 - 2)$. To perform this subtraction we subtract 1 from $a_3$, add $R - 1 = 6$ to $a_2$, and $d_1 = (a_1 + R) - b_1 = 1 + 7 - 2 = 6$ is the second output digit. The third digit subtraction is now $(6 - 6) = 0 = d_2$ and the fourth digit subtraction is $(5 - 4) = 1 = d_3$. Thus $6016_7 - 4625_7 = 1061_7$. The subtraction can be represented in the following format.

$$
\begin{array}{ll}
\phantom{-}6016 & 1^{\text{st}} \text{ number} \\
\underline{-4625} & 2^{\text{nd}} \text{ number} \\
\phantom{-}\phantom{6016}1 & \text{borrow} \\
\underline{\phantom{-}\phantom{60}67} & \text{lend} \\
\phantom{-}1061 & \text{result}
\end{array}
$$

If $B > A$, $A - B$ is a negative number. In this case Procedure 2.4 will not lead to a meaningful result. To rectify this, we must first define a means of representing negative numbers. It is possible to represent negative numbers in such a manner that subtraction can be performed via the use of the addition algorithm. This is significant from a digital system design viewpoint since it implies that the same circuit can be used for both addition and subtraction. In order to understand this representation we must first introduce the concept of *modulo number representation*.

For any pair of integers $x,y$ where $y > 0$ the number $x$ has a unique representation of the form $x = k_1 \cdot y + k_2$ where $k_1, k_2$ are integers and $0 \le k_2 < y$. For example if $y = 6$ and $x = 28$, $x = 4 \cdot 6 + 4$. The number $k_2$ in this representation of $x$ in terms of $y$ is referred to as $x$ modulo $y$ or $x$ mod $y$. If $x = -11$ and $y = 5$, $x = (-3) \cdot 5 + 4$ and hence $-11 \bmod 5 = 4$.

**Lemma 2.1**

If $x,y$ and $k$ are integers then $(x + k \cdot y) \bmod y = x \bmod y$

**Proof**

Assume $x \bmod y = k_1$. Then $x = k_2 \cdot y + k_1$. Therefore

$$x + k \cdot y = (k_2 \cdot y + k_1) + k \cdot y$$
$$= (k_2 + k) \cdot y + k_1$$
$$= k' \cdot y + k_1 \quad \text{where } k' = (k_2 + k)$$

Therefore $(x + k \cdot y) \bmod y = k_1 = x \bmod y$

In performing addition the sum of two numbers may be greater than either of the numbers. If the two operands $A,B$ in an addition are represented as $k$-digit base $R$ numbers, then $A,B \leq R^k - 1$ and the sum $A + B \leq 2R^k - 2$ which may require a $(k + 1)$-digit representation. If only the $k$ least significant digits are used to represent $A + B$ the resulting number is $A + B \bmod R^k$ since if $A + B > R^k - 1$ the most significant digit, which was deleted, is 1 and the result is $A + B - R^k = (A + B) \bmod R^k$. For instance, assuming 4-digit representation, if $A = 1001_2$ and $B = 0111_2$, the addition of $A$ and $B$ yields

$$
\begin{array}{l}
\quad 1001 \\
\quad 0111 \\
\underline{\quad 1111} \quad \text{carries} \\
1\underbrace{0000}
\end{array}
$$

4-bit representation

Deleting the most significant digit of $A + B$ the result becomes 0000. This is because $A + B = 16$ and $(16) \bmod 2^4 = (16) \bmod 16 = 0$. Thus the operation of addition when performed with $k$-digit operands and $k$-digit result is actually modulo $R^k$ addition.

The operation of subtraction $A - B$, for $A,B < R^k$ can be expressed as an addition operation involving a negative operand, $A + (-B)$. Assuming addition is modulo $R^k$ then $(A + (-B)) \bmod R^k = [(A + (-B)) + R^k] \bmod R^k = [A + (R^k - B)] \bmod R^k$. Thus the modulo $R^k$ addition can be performed with two positive operands, $A$ and $R^k - B$. $R^k - B$ is called the $R$-complement representation of $-B$ (for $R = 2$, it is called the 2's complement). A number can be converted to 2's complement form without using subtraction. Assuming $R = 2$, $R^k - B = 2^k - B$ which can be represented as $(2^k - 1) - B + 1$.

The number $2^k - 1$ is represented as $\underbrace{111...1}_{k \text{ times}}$. Hence $(2^k - 1) - B$ can be generated by simply changing all 0 bits of $B$ to 1 and all 1 bits of $B$ to 0. The resulting number is called the *complement of B*. To form $2^k - B$ we thus complement $B$ and then add 1. Similarly, the $R$-complement of a base $R$ number $B = b_{k-1}...b_1 b_o$ is computed by changing $B_i$ to $(R - 1 - B_i)$, $0 \le i \le k - 1$, and then adding 1.

### Procedure 2.5

To compute the $R$ complement of a base $R$ number $B = b_{k-1}...b_1 b_o$
(1) change all digits $b_i$ to $R - 1 - b_i$ for $0 \le i < k$. (2) Then add 1 to the resulting number.

### Example 2.7

The 2's complement representation of $-13$ is formed by complementing the binary representation of 13, 01101, resulting in 10010, and adding 1 resulting in 10011. Note that adding first and then complementing yields the incorrect result of 10001.

The 2's complement of a number $B$ can also be obtained by complementing all bits to the left of the least significant (rightmost) 1 bit in $B$. (See Problem 2.14). For example if $B = 011001$ we complement the leftmost 5 bits to form the 2's complement of $B$ which is 100111. Similarly if the number is 10011100 the 2's complement is 01100100.

In order to ensure that $A + B < R^k$ we assume that $A,B < R^{k-1}$. If $B < 2^{k-1}$, then the 2's complement of $B$ will always have a 1 in bit $b_k$. We will therefore use this bit as a sign bit. A sign bit equal to 1 indicates a negative number and a sign bit equal to 0 indicates a positive number. Given a 2's complement representation of a number $B$, $B' = 2^k - B$, the 2's complement of $B'$ is equal to $2^k - (2^k - B) = B$. Thus if $k = 4$, the number 10110 represents a negative number, since the most significant bit is 1. Taking the 2's complement of this number we obtain 01010 which is the binary representation of 10. Thus 10110 is the 2's complement representation of $-10$. Subtraction can now be performed by converting all negative operands to 2's complement representation and then performing addition. In order that we obtain the correct result, mod $2^k$, each of the operands $A,B$ is restricted to be within the range of values $-2^{k-1} < A,B < 2^{k-1}$. If the sign bit addition is performed as specified in the following procedure the sign bit of the result will always be correct.

**Procedure 2.6 (Addition/Subtraction of Two Operands)**
(1) Convert any negative operands to 2's complement form.
(2) Add the operands as specified in Procedure 2.3 treating the sign bit in the same manner as all other bits and permitting a carry into the (high order) sign bit position, while ignoring a carry out of the sign bit position into bit position $k + 1$.

It remains to be proven that the sign bit of the result is always correct.

**Theorem 2.1**
If addition and subtraction are performed as specified in Procedure 2.6, the result sign bit will be 0 if the result is $\geq 0$ and the sign bit will be 1 if the result is $< 0$.

**Proof**
We must consider three distinct cases, corresponding to addition of two positive numbers, one positive and one negative number, and two negative numbers.

Case #1: Assume both numbers are non-negative. Then if these numbers are $A = a_{k-1}...a_1 a_o$ and $B = b_{k-1}...b_1 b_o$ they are represented as $0 a_{k-1}...a_1 a_o$ and $0 b_{k-1}...b_1 b_o$. Assuming the addition is such that $A + B \leq 2^k - 1$ (which is required for modulo $2^k$ addition) then there will be no carry into the sign bit position and therefore the result sign bit is 0, which is correct since $A + B \geq 0$.

Case #2: Assume $A$ is non-negative and $B$ is negative. Then $A$ is represented as $0\ B_A$ (where $B_A$ is the binary representation of $A$) and $B$ is represented as $1\ B_C$ where $B_C$ is the binary representation of $C = 2^k - B$. If $A \geq B$ then this addition will result in a carry into the sign position since $B_C + B_A = 2^k - B + A \geq 2^k$. Therefore the result sign bit will be $0 + 1 + 1 = 0 \mod 2$ which is correct since if $A \geq B$, $A - B \geq 0$. If $A < B$ then $B_C + B_A = 2^k - B + A < 2^k$ and there is no carry into the sign position. The result sign bit will be $0 + 1 + 0 = 1$ which is correct since if $A < B$, $A - B < 0$.

Case #3: Assume both $A$ and $B$ are negative. Then $A$ is represented as $1\ B_D$ where $D = 2^k - A$ and $B$ is represented as $1\ B_C$. Then $B_D + B_C = (2^k - A) + (2^k - B) = 2^{k+1} - (A + B)$. But $-(A + B)$ is restricted to be $\geq -(2^k - 1)$ for correct modulo $2^k$ addition and hence $B_D + B_C \geq 2^k$. Therefore there will be a carry into the sign position and the sign bit of the result will be $1 + 1 + 1 = 1 \mod 2$ which is correct since $A + B < 0$.

**Example 2.8**

a) Consider the addition 4 + 6 (where $k = 4$). Since neither operand is negative, we add the corresponding binary numbers with 0 in the sign bit position

```
   sign bit
  /
0 0 1 0 0
0 0 1 1 0
    1              Carry
─────────
0 1 0 1 0
```

The result is +10 (since the sign bit is 0 and the magnitude bits are 1010).

b) Consider the addition 6 + (−4). The negative number is represented by its two complement, 11100. The addition is:

```
   sign bit
  /
0 0 1 1 0
1 1 1 0 0
1 1                Carry
─────────
0 0 0 1 0
  \
   sign bit
```

The result is +2. (Note the carry into the sign bit position.)

c) Consider the addition −6 + 4. This addition is represented as:

```
   sign bit
  /
1 1 0 1 0    (2's complement representation of −6)
0 0 1 0 0
─────────
1 1 1 1 0
  \
   sign bit
```

The result is a negative number, since the sign bit is 1. The number can be determined by taking the 2's complement of the result yielding 00010. Thus the result is $-2$.

d) Consider the addition $-6 + (-4)$. Both numbers are represented in 2's complement form and added as follows:

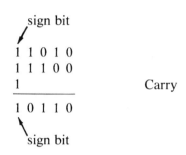

The result is negative. The 2's complement of the result is 01010. Thus the result is $-10$.

Procedure 2.6 can be generalized to perform base $R$ subtraction using addition, where the $R$-complement of $B$ is generated as specified in Procedure 2.5.

The *one's complement* representation of a negative number $B$, defined as $2^k - B - 1$, can be obtained by simply complementing each bit of $B$. Procedure 2.6 can be modified so as to perform binary addition and subtraction using one's complement representation of negative numbers (Problem 2.16). A negative result can be converted to the corresponding binary number simply by complementing each bit of the result.

### 2.2.3 BASE $R$ MULTIPLICATION

Consider the base $R$ multiplication of two numbers $A = a_n a_{n-1} \ldots a_1 a_0$ and $B = b_n b_{n-1} \ldots b_1 b_0$.

$$A \cdot B = \left( \sum_{i=0}^{n} a_i \cdot R^i \right) \cdot B$$

$$= a_0 \cdot B + a_1 \cdot RB + a_2 \cdot R^2 B + \ldots + a_n \cdot R^n B$$

Furthermore    if    $Y = y_n y_{n-1} \ldots y_1 y_0$    the    product    $R^i \cdot Y =$

$y_n y_{n-1} \cdots y_1 y_o \underbrace{00...0}_{i \text{ zeros}}$ . (Exercise). Thus the multiplication of $A \cdot B$ can be expressed as a sum of $n$ much simpler multiplications of the form $a_i \cdot (y_n y_{n-1} \cdots y_1 y_o \underbrace{00...0}_{i \text{ zeros}})$. For $R = 2$, $a_i = 0$ or 1 and $0 \cdot B \cdot R^i = 0$

and $1 \cdot B \cdot R^i = B \cdot R^i$. Hence binary multiplication is especially simple. It consists of adding all terms of the form $2^i \cdot B$ for all $i$ such that $a_i = 1$.

**Procedure 2.7 (Binary Multiplication)**
To multiply $A = a_n a_{n-1} \cdots a_1 a_o$, by $B = b_n b_{n-1} \cdots b_1 b_o$
1) Set $Z = 0$, $i = 0$.
2) If $a_i = 1$ add $2^i B = (b_n b_{n-1} \cdots b_o \underbrace{0...0}_{i})$ to $Z$
3) Increment $i$ by 1
4) Repeat 2) and 3) for all $i$, $0 \le i \le n$. At compĺetion $Z$ contains the binary representation of $A \cdot B$. If $A$ and $B$ are $n$-bit numbers, $A \cdot B$ has at most $2n$ bits.

**Example 2.9**
Let $A = 01011$ and $B = 11000$. Initially $Z = 0$. Since $a_o = 1$ we add $B$ to $Z$ resulting in $Z = 11000$. Since $a_1 = 1$ we add $2B = (110000)$ to $Z$ resulting in $Z = (1001000)$. Since $a_3 = 1$ we add $2^3 \cdot B = (11000000)$ to $Z$ resulting in $Z = (100001000) = A \cdot B$. Multiplication can be represented as follows

$$
\begin{array}{rl}
B = & 1\ 1\ 0\ 0\ 0 \\
A = & 0\ 1\ 0\ 1\ 1 \\
\hline
 & 1\ 1\ 0\ 0\ 0 = 2^o \cdot B \\
 & 1\ 1\ 0\ 0\ 0\ 0 = 2^1 \cdot B \\
 & 1\ 1\ 0\ 0\ 0\ 0\ 0\ 0 = 2^3 \cdot B \\
\hline
 & 1\ 1\ 1\ 1 \quad \text{Carries} \\
\hline
A \cdot B = & 1\ 0\ 0\ 0\ 0\ 1\ 0\ 0\ 0
\end{array}
$$

**2.2.4 BASE $R$ DIVISION**

The operation of division is somewhat more complex due to the fact that for two integers $A$, $B$, $A \div B$ may not be an integer. Suppose in base $R$,

$$A \div B = \sum_{i=0}^{n} C_i R^i + \frac{R'}{B}$$

where $R'$, the remainder, is such that $0 \le R' < B$.

Then
$$A = B \cdot \sum_{i=0}^{n} C_i R^i + R'$$

The digit $C_n$ must satisfy the constraint

$$A - B \cdot R^n \cdot C_n = B \cdot \sum_{i=0}^{n-1} C_i R^i + R'$$

Thus
$$B \cdot (R^n - 1) > (A - C_n R^n B) \ge 0.$$

This uniquely defines $C_n$. The remaining digits $C_{n-1}, C_{n-2} \ldots C_o$ are defined similarly, most significant digit first, by constraints of the form

$$B \cdot (R^i - 1) > (A' - C_i R^i B) \ge 0, \text{ where}$$

$$A' = \left( A - B \cdot \sum_{j=i+1}^{n} C_j R^j \right)$$

For binary division, the process is again simplified. In this case $C_n$ = 1 if $A - 2^n B = A - b_n b_{n-1} \ldots b_1 b_0 \underbrace{0 \ldots 0}_{n} \ge 0$, otherwise $C_n$ = 0. If $C_n = 1$, the remaining bits $C_{n-1} \ldots C_o$ must be such that $A = 2^n \cdot B + \sum_{i=0}^{n-1} C_i \cdot 2^i \cdot B$. These remaining bits are generated by first subtracting $2^n \cdot B$ from $A$ resulting in $A' = A - 2^n B$ which is then used to define the succeeding bits $C_{n-1}, C_{n-2} \ldots C_o$. Thus division is effectively accomplished by repeated subtraction.

### Procedure 2.8 (Binary Division)

1) Set $i = n$, the remainder $R' = A$.
2) If $R' - 2^i B \ge 0$ then set $C_i = 1$ and subtract $2^i \cdot B$ from $A'$. If $R' - 2^i \cdot B < 0$ then set $C_i = 0$.
3) Repeat (2) for $i = n - 1, n - 2, \ldots 1, 0$ to generate the bits $C_{n-1}, C_{n-2}, \ldots, C_1, C_o$. At completion the binary representation of $A \div B$ is $C_n C_{n-1} \ldots C_1 C_o$ and the remainder is $R'$.

**Example 2.10**

We will divide $A = 11101_2$ by $B = 00101_2$. The result $C = C_4 C_3 C_2 C_1 C_0$ is obtained as follows: Since $A - 2^4 B = A - 001010000 < 0$, $C_4 = 0$. Similarly $C_3 = 0$ since $A - 2^3 B < 0$. However, $A - 2^2 B > 0$. Therefore $C_2 = 1$ and $A' = A - 2^2 B = 11101 - 10100 = 01001$. $A' - 2^1 \cdot B = 01001 - 01010 < 0$. Therefore $C_1 = 0$. Finally $A' - 2^0 \cdot B = 01001 - 00101 = 00100$. Thus $C_0 = 1$. The binary representation of $A \div B$ is $C = 00101$ and the remainder is $R' = 00100$. This division can be represented as follows:

$$
\begin{array}{r}
00101 \\
00101\overline{\big)\,11101\phantom{00}} \\
-10100\,(2^2 \cdot B) \\
\hline
01001 \\
-00101\,(2^0 \cdot B) \\
\hline
00100
\end{array}
$$

In succeeding chapters we shall design digital circuits and systems which perform binary arithmetic computations as specified in the algorithms presented in this section.

### 2.2.5 FLOATING POINT ARITHMETIC

Until now we have limited our discussion to *fixed point representation* of numbers in which all of the digits of a number must be represented. In many applications the numbers to be represented may require many digits. These numbers can frequently be represented with fewer digits using *floating point notation*. For example the number 5,000,000 can be represented as $5 \times 10^6$. In digital systems, binary floating point numbers may be represented in the *normalized* form $M \cdot 2^{\pm E}$ where $M$, the *mantissa*, is constrained to be, $1/2 \le M < 1$ and $E$ is the *exponent* of the number, and both $M$ and $E$ are represented in binary. Thus 1011000 would be represented as $(.1011) \cdot 2^{111}$ and .00001 would be represented as $(.1) \cdot 2^{-100}$. In multiplication or division, exponents are respectively either added or subtracted.

Thus if $$X = M_x \cdot 2^{E_x}$$

and $$Y = M_y \cdot 2^{E_y}$$

then $$X \cdot Y = M_x \cdot M_y \cdot 2^{E_x + E_y}$$

and $$X \div Y = (M_x \div M_y) \cdot 2^{E_x - E_y}$$

If $M_x \cdot M_y$ or $M_x \div M_y$ is not within the normalization constraints defined on the mantissa then, in order to normalize, the result must be modified by changing the exponent. For example

$$(.1)2^1 \cdot (.1)2^1 = (.01) \cdot 2^{10} = (.1) \cdot 2^1.$$

To perform addition or subtraction on floating point numbers the numbers must have the same exponent. (This corresponds to proper alignment of the numbers). If $X = M_x \cdot 2^{E_x}$ and $Y = M_y \cdot 2^{E_y}$ and $E_x = E_y$ then $X + Y = (M_x + M_y) \cdot 2^{E_x}$. Thus the addition $(.1)2^1 + (.1)2^{10}$ is performed as $(.01)2^{10} + (.1)2^{10} = (.11)2^{10}$. Normalization may again be required.

## 2.3 OTHER CODES

The two number codes we have considered, the binary code (i.e. binary number system) and the BCD code are examples of *weighted* codes in which the code number can be interpreted by assigning a weight

| Number | Excess-3 Representation | | | |
|--------|---|---|---|---|
| 0 | 0 | 0 | 1 | 1 |
| 1 | 0 | 1 | 0 | 0 |
| 2 | 0 | 1 | 0 | 1 |
| 3 | 0 | 1 | 1 | 0 |
| 4 | 0 | 1 | 1 | 1 |
| 5 | 1 | 0 | 0 | 0 |
| 6 | 1 | 0 | 0 | 1 |
| 7 | 1 | 0 | 1 | 0 |
| 8 | 1 | 0 | 1 | 1 |
| 9 | 1 | 1 | 0 | 0 |

Figure 2.2   Excess-3 Code

to each bit position and adding those positions in which 1's occur, or each digit of the number can be interpreted by applying this procedure to sets of 4 consecutive bits, respectively. However other types of codes have proven useful in various applications. One such code is the *Excess-3 code*. In this code each digit $D$ is represented by the binary encoding of $(D + 3)$. This code, which is shown in Figure 2.2 has the property that the 9's complement of any number can be obtained by complementing each bit of the number. This type of code which is called a *self-complementing code* was prevalently used in early decimal computers.

In some applications it is desirable to have a $k$-bit representation of the numbers 0 to $2^k - 1$ in which all successive code words differ in only one variable as do the codes for $2^k - 1$ and 0. Such a code is called a cyclic code. The *Gray code* is one such code which has had applications in design of counting circuits. The 2-bit and 3-bit Gray codes are shown in Figure 2.3.

| | $x_1$ | $x_2$ | | | $x_1$ | $x_2$ | $x_3$ |
|---|---|---|---|---|---|---|---|
| 0 | 0 | 0 | | 0 | 0 | 0 | 0 |
| 1 | 0 | 1 | | 1 | 0 | 0 | 1 |
| 2 | 1 | 1 | | 2 | 0 | 1 | 1 |
| 3 | 1 | 0 | | 3 | 0 | 1 | 0 |
| | | | | 4 | 1 | 1 | 0 |
| | (a) | | | 5 | 1 | 1 | 1 |
| | | | | 6 | 1 | 0 | 1 |
| | | | | 7 | 1 | 0 | 0 |

(b)

Figure 2.3 (a) Two bit Gray Code (b) Three bit Gray Code

A $k$-bit Gray code $G$ can be obtained from a $(k - 1)$-bit Gray code $G'$ as shown in Figure 2.4, by adding a bit $x_1$, which is 0 for the numbers 0 through $(2^{k-1} - 1)$, and the remaining bits of the code are identical to $G'$ for these numbers. The remaining numbers $(2^{k-1}$ through $(2^k - 1))$ have $x_1 = 1$ and the other bits are defined by $G'_r$ which is obtained by reversing the rows of $G$, first row becomes last and vice versa.

Other codes have been developed which simplify the problem of

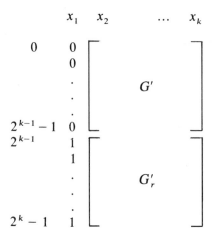

$$
\begin{array}{cc}
 & x_1 \quad x_2 \quad \cdots \quad x_k \\
0 & 0 \\
 & 0 \\
 & \cdot \\
 & \cdot \\
 & \cdot \\
2^{k-1}-1 & 0 \\
2^{k-1} & 1 \\
 & 1 \\
 & \cdot \\
 & \cdot \\
 & \cdot \\
2^k-1 & 1
\end{array}
$$

Figure 2.4   A k-bit Gray Code

detecting errors in a digital system. A *parity check code* is an example of such a code. Given any $k$-bit word in a digital system we convert it to a $(k + 1)$-bit code by defining the extra bit in such a manner that the total number of 1-bits in the $(k + 1)$-bit word is even.* Now an error in a single bit of a word causes it to have an odd number of 1-bits and hence this error can be detected. Such a code is called an *error detecting code*. An *error correcting code* enables the determination of which bit is in error. Numerous codes have been developed to enhance the reliability of different types of digital systems.

**SOURCES**

The subject of number systems is considered in almost all elementary books on digital computers. An extensive presentation of binary arithmetic procedures is presented by Flores[1]. The material on error detecting and correcting codes is originally due to Hamming[2] and an extensive survey of other such codes and their properties is contained in a book by Sellers, Hsiao and Bearnson[3].

---

*Alternately an *odd-parity* code can be defined in which each word has an odd number of 1-bits

## REFERENCES

[1]  Flores, I. *The Logic of Computer Arithmetic*, Prentice-Hall, Inc. Englewood Cliffs, N.J., 1963.

[2]  Hamming, R. W. "Error Detecting and Error Correcting Codes," *Bell System Technical Journal*, vol. 29, pp 147-160, April, 1950

[3]  Sellers, F. F. Jr., Hsiao, M. Y. and Bearnson, L. W., *Error Detecting Logic for Digital Computers*, McGraw Hill, New York, NY, 1968

## PROBLEMS

**2.1)** Convert the decimal number 751.3 to each of the bases 2, 3, 7, 11. In each case expand the fractional part of the number to a sufficient number of positions so that the maximum error does not exceed $.02_3$.

**2.2)** Do each of the following arithmetic operations in the indicated number base.
a) $2345_9 + 16250_9$
b) $324_5 \cdot 23_5$
c) $651_7 \div 46_7$
d) $1101_2 \cdot 101_2$

**2.3)** Do the following subtraction operations using $r$-complement addition.
a) $11011_2 - 00101_2$
b) $6562_7 - 5616_7$

**2.4)** Solve the following set of simultaneous equations for $x$ and $y$.
$40_y = 44_6$
$x1_y = 111_x$

**2.5)** Convert the following numbers to base 2.
a) $614_7$
b) $43.4_5$
c) $1(10)_{11}$
d) $110_{11}$

**2.6)** Convert the following numbers to base 5.
   a) $101110.011_2$
   b) $614_9$

**2.7)** The following multiplication is done in base $y$, $0 < y < 10$. Find all values of $y$ for which the multiplication is valid.
   a) $21_y \cdot 12_y = 1022_y$
   b) $21_y \cdot 12_y = 253_y$

**2.8)** The following addition is done in base 5. Find $a,b,c$, $0 \le a,b,c \le 4$.

$$aaba_5 + bac1_5 = 1ab00_5.$$

**2.9)** Perform the following operations in binary.
   a) $10 - 5$ using 2's complement addition
   b) $5 - 10$ using 2's complement addition
   c) $-10 - 5$ using 2's complement addition
   d) $10 \cdot 13$
   e) $39 \div 6$

**2.10)** For each of the following operations determine in which number bases it is valid.
   a) $1234 + 5432 = 6666$
   b) $\sqrt{41} = 5$
   c) $111_x = x1_{(x+1)}$

**2.11)** If $A = .101 \cdot 2^{101}$ and $B = .11 \cdot 2^{-11}$ find $A + B$, $A - B$, and $A \cdot B$. Express your answer in normalized floating point form.

**2.12)** Assume that a digital system has all registers having $n$ bits and addition and subtraction is done in 2's complement form. If the result $R$ is not in the range $-2^n < R < 2^n$, then $R$ cannot be stored in a $n$-bit register. In this case an *overflow error* is said to have occurred. Specify how to detect such an error.

**2.13)** From the Gray codes of Figure 2.3, derive 4 and 5-bit Gray codes.

**2.14)** a) Prove that the 2's complement of a binary number $N$ can be formed by complementing all bits to the left of the least significant 1 bit.

b) Derive a similar procedure for forming the $r$-complement of a base $r$ number.

**2.15)** Assume that the words of a digital system are encoded in an even parity check code of 4 bits. Which of the outputs 0011, 1011, 1110, if received would indicate an error.

**2.16)** Modify Procedure 2.6 to perform one's complement addition and subtraction and use this procedure to perform the computations specified in Example 2.8.

# part 2

# Combinational and Sequential Circuits

# CHAPTER 3

# Combinational Circuit Synthesis

## 3.1 FUNCTION SPECIFICATION

For a Boolean combinational function, $f(x_1, x_2, ..., x_n)$, the binary input variables $x_i$ can each take the value 0 or 1, so there are $2^n$ possible values of the set of variables $x_1, x_2, ..., x_n$. These $2^n$ possible values define an *n-dimensional space or cube*, and each specific value within this cube is referred to as a *point*. For any of the $2^n$ points $x_i$* within the cube, $f(x_i)$ is equal to 1 (in which case $x_i$ is called a *1-point* of f), or $f(x_i)$ is equal to 0 ($x_i$ is a *0-point of f*) or $f(x_i)$ is unspecified ($x_i$ is a *don't care point of f*). An unspecified value for a particular input state is used to indicate that that input state will never occur or that the value of the function for that input state can be defined to be either 0 or 1. A combinational function f can be specified by a truth table which has $2^n$ rows, one for each possible input state. Associated with each such row is the corresponding value of f, where unspecified entries are denoted by -'s in the table. The truth table of Figure 3.1 specifies three functions $f_1, f_2, f_3$, all of which have input variables $x_1, x_2, x_3$. For input state $x_i = (0,1,0)$ (i.e. $x_1 = 0$, $x_2 = 1$, $x_3 = 0$), $f_1(x_i) = 1$, $f_2(x_i)$ is unspecified, and $f_3(x_i) = 1$.

In this chapter we will consider the problem of deriving a digital circuit to realize a combinational function specified by a word description. The first step in this *synthesis* problem is frequently the derivation of

---

*The notation $x_i$ is used to denote a specific value $(a_{i_1}, a_{i_2}, ...a_{i_n})$, $a_{ij} = 0$ or 1 for all $j$ of a set of binary variables $(x_1, x_2, ...x_n)$, and may be referred to as an *input state* or *input combination*. The set of variables will be denoted as x and thus $f(x_1, x_2, ...x_n)$ may be represented as $f(\mathbf{x})$.

| $x_1$ | $x_2$ | $x_3$ | $f_1$ | $f_2$ | $f_3$ |
|-------|-------|-------|-------|-------|-------|
| 0     | 0     | 0     | 1     | 1     | 1     |
| 0     | 0     | 1     | 0     | 0     | 0     |
| 0     | 1     | 0     | 1     | -     | 1     |
| 0     | 1     | 1     | 0     | -     | 0     |
| 1     | 0     | 0     | 1     | 1     | -     |
| 1     | 0     | 1     | 1     | 1     | 0     |
| 1     | 1     | 0     | 0     | 0     | 1     |
| 1     | 1     | 1     | 0     | 0     | 0     |

Figure 3.1    A truth table of combinational functions

a truth table describing the function to be realized. (This is not *always* done as we shall demonstrate in Chapter 4.) The derivation of the truth table may involve the necessity of formulating a binary representation of the inputs and outputs of the function, and can usually be formulated as the following 3-step process.

(1) Represent the function inputs as binary valued variables $x_i$.
(2) Represent the function output as binary valued variables $z_i$.
(3) For each input combination, specify the value each output should assume.

Although there exists no systematic procedure to generate a truth table (due to the vague concept of what constitutes a word description of a function) it can usually be accomplished without great difficulty as illustrated in the following examples.

**Example 3.1**

We will specify a truth table for the function $f(\mathbf{x}) = \lceil \sqrt{x} \rceil$ where $\lceil y \rceil$ is the smallest integer $\geq y$ and $0 \leq \mathbf{x} \leq 15$. The input $\mathbf{x}$ will be represented by 4 binary variables, $x_1, x_2, x_3, x_4$ which are to be interpreted as a binary number $\mathbf{x} = x_1 x_2 x_3 x_4$. The output of the circuit is constrained within the range $0 \leq f(\mathbf{x}) \leq 4$. The output will be represented by 3 binary variables $z_1, z_2, z_3$ which are to be interpreted as a binary encoded number $\mathbf{z} = z_1 z_2 z_3$. The truth table (Figure 3.2) lists all 16 possible values of $x_1, x_2, x_3, x_4$, the corresponding value of $\mathbf{x}$ and $\mathbf{z}$, and the binary outputs $z_1, z_2, z_3$ which are interpreted as the binary encoding of $\mathbf{z}$.

| x | z = f(x) | $x_1$ | $x_2$ | $x_3$ | $x_4$ | $z_1$ | $z_2$ | $z_3$ |
|---|---|---|---|---|---|---|---|---|
| 0 | 0 | 0 | 0 | 0 | 0 | 0 | 0 | 0 |
| 1 | 1 | 0 | 0 | 0 | 1 | 0 | 0 | 1 |
| 2 | 2 | 0 | 0 | 1 | 0 | 0 | 1 | 0 |
| 3 | 2 | 0 | 0 | 1 | 1 | 0 | 1 | 0 |
| 4 | 2 | 0 | 1 | 0 | 0 | 0 | 1 | 0 |
| 5 | 3 | 0 | 1 | 0 | 1 | 0 | 1 | 1 |
| 6 | 3 | 0 | 1 | 1 | 0 | 0 | 1 | 1 |
| 7 | 3 | 0 | 1 | 1 | 1 | 0 | 1 | 1 |
| 8 | 3 | 1 | 0 | 0 | 0 | 0 | 1 | 1 |
| 9 | 3 | 1 | 0 | 0 | 1 | 0 | 1 | 1 |
| 10 | 4 | 1 | 0 | 1 | 0 | 1 | 0 | 0 |
| 11 | 4 | 1 | 0 | 1 | 1 | 1 | 0 | 0 |
| 12 | 4 | 1 | 1 | 0 | 0 | 1 | 0 | 0 |
| 13 | 4 | 1 | 1 | 0 | 1 | 1 | 0 | 0 |
| 14 | 4 | 1 | 1 | 1 | 0 | 1 | 0 | 0 |
| 15 | 4 | 1 | 1 | 1 | 1 | 1 | 0 | 0 |

Figure 3.2  Truth table for squareroot function

## Example 3.2

We will specify a truth table for a circuit which is intended to be a candy dispenser and change maker for a simple candy vending machine. The machine will have two types of candy, $c_1$ which costs 10 cents and $c_2$ which costs 25 cents. Candy $c_1$ can be purchased by putting a dime in coin slot #1 in the machine and pushing the button $c_1$, or by putting a quarter in the machine in coin slot #2 and pushing button $c_1$, resulting in change in coin slot #3 of 1 nickel and in coin slot #4 of 1 dime. Candy $c_2$ can only be purchased by putting a quarter in coin slot #2 and pushing button $c_2$.

The function has the following binary inputs:

$x_d$ = 1 if a dime is placed in coin slot #1
    = 0 otherwise
$x_q$ = 1 if a quarter is placed in a coin slot #2
    = 0 otherwise
$x_{c_1}$ = 1 if button $c_1$ is pushed
    = 0 otherwise

$x_{c_2}$  $= 1$ if button $c_2$ is pushed
$\phantom{x_{c_2}}$ $= 0$ otherwise

and the following binary outputs:

$z_n$  $= 1$ if a nickel is to be returned as change in coin slot #3
$\phantom{z_n}$  $= 0$ otherwise
$z_d$  $= 1$ if a dime is to be returned as change in coin slot #4
$\phantom{z_d}$  $= 0$ otherwise
$z_{c_1}$  $= 1$ if candy $c_1$ is to be dispensed
$\phantom{z_{c_1}}$  $= 0$ otherwise
$z_{c_2}$  $= 1$ if candy $c_2$ is to be dispensed
$\phantom{z_{c_2}}$  $= 0$ otherwise

The truth table (Figure 3.3) lists the 16 possible values of the input variables $(x_d, x_q, x_{c_1}, x_{c_2})$ and for each of these the associated values of the output functions $(z_n, z_d, z_{c_1}, z_{c_2})$.

The outputs for some of the input states in the table have not been

| Row | $x_d$ | $x_q$ | $x_{c_1}$ | $x_{c_2}$ | $z_n$ | $z_d$ | $z_{c_1}$ | $z_{c_2}$ |
|-----|-------|-------|-----------|-----------|-------|-------|-----------|-----------|
| 1 | 0 | 0 | 0 | 0 | 0 | 0 | 0 | 0 |
| 2 | 0 | 0 | 0 | 1 | 0 | 0 | 0 | 0 |
| 3 | 0 | 0 | 1 | 0 | 0 | 0 | 0 | 0 |
| 4 | 0 | 0 | 1 | 1 | 0 | 0 | 0 | 0 |
| 5 | 0 | 1 | 0 | 0 | 0 | 0 | 0 | 0 |
| 6 | 0 | 1 | 0 | 1 | 0 | 0 | 0 | 1 |
| 7 | 0 | 1 | 1 | 0 | 1 | 1 | 1 | 0 |
| 8 | 0 | 1 | 1 | 1 | 0 | 0 | 0 | 1 |
| 9 | 1 | 0 | 0 | 0 | 0 | 0 | 0 | 0 |
| 10 | 1 | 0 | 0 | 1 | 0 | 1 | 0 | 0 |
| 11 | 1 | 0 | 1 | 0 | 0 | 0 | 1 | 0 |
| 12 | 1 | 0 | 1 | 1 | 0 | 0 | 1 | 0 |
| 13 | 1 | 1 | 0 | 0 | 0 | 0 | 0 | 0 |
| 14 | 1 | 1 | 0 | 1 | 0 | 1 | 0 | 1 |
| 15 | 1 | 1 | 1 | 0 | 1 | 1 | 1 | 0 |
| 16 | 1 | 1 | 1 | 1 | 0 | 0 | 1 | 1 |

**Figure 3.3    Truth table for candy and charge dispenser**

explicitly stated in the word description but have been specified in the truth table in what is considered to be a reasonable manner. Thus for rows 1-4, in which $x_d = x_q = 0$ (no money put in machine) the outputs are all 0. For rows 5, 9, 13 where $x_{c_1} = x_{c_2} = 0$ (no candy selection button has been pushed) the outputs are also 0. Rows 6, 7, 11 constitute proper operation as specified in the word description. Rows 8, 12 in which both buttons are pushed dispenses the candy paid for. Row 10 in which a dime has been paid and $c_2$ selected, returns the dime. Finally rows 14, 15, 16 in which both a dime and a quarter have been paid dispenses the candy (or candies) selected and the correct change for row 14, and the maximum permissible change for row 15. Note that if we were to assume that the machine would never have an improper input (rows 1, 6, 7, 11 being the only proper inputs) then the remaining rows could have unspecified outputs. In this case $z_n = z_d$. (Change always consists of 1 nickel and 1 dime.)

Once the truth table has been derived we then wish to design a circuit to realize the binary function (or functions) specified by the table. A circuit is said to *realize* a binary combinational function $f$ if the output of the circuit is 1 for any input corresponding to a 1-point, and the output is 0 for any input corresponding to a 0-point. The output corresponding to a don't care point can be either 0 or 1.* The circuit of Figure 3.4 with output $F(x_1,x_2,x_3)$ realizes $f_1$ of Figure 3.1 (since $F(x_1,x_2,x_3) = f_1(x_1,x_2,x_3)$) and also $f_2$ (since $f_2(x_1,x_2,x_3)$ and $F(x_1, x_2,x_3)$ are equal for all input states for which $f_2$ is specified). However the circuit does not realize $f_3$ since $f_3(1,0,1) = 0$ and $F(1,0,1) = 1$.

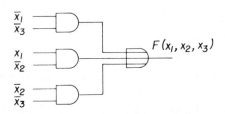

Figure 3.4    A combinational logic circuit

---

*Although a combinational function may be unspecified for an input state **x**, the output of any digital circuit is always 0 or 1 for any input state. Thus the output of a combinational circuit always corresponds to a completely specified combinational function.

For an $n$-dimensional cube defined by variables $(x_1, x_2, \ldots, x_n)$ each point can be represented as a product term in which each of the $n$ variables appears complemented or uncomplemented. For example, if $n = 3$ the product term $x_1 \bar{x}_2 x_3$ represents the point $(x_1, x_2, x_3) = (1,0,1)$. Such a product term is called a *minterm*, since it defines a Boolean function which has the value 1 for only one point in the $n$-dimensional space. It is convenient to label minterms as $m_i$ where $i$ is the integer defined by treating the product term as a binary number sequence of 0's (for each complemented variable) and 1's (for each uncomplemented variable) where $x_1$ defines the most significant bit. Thus $x_1 \bar{x}_2 x_3$ defines the number 101 which represents $m_5$, since 101 is a binary representation of the integer 5. Similarly $m_3$ corresponds to $\bar{x}_1 x_2 x_3$ which defines the binary number 011.

A combinational function $f(x_1, \ldots, x_n)$ can be realized by a circuit of the form shown in Figure 3.5. This circuit has several $n$-input AND gates and one OR gate. The AND gates are used to realize the minterms which correspond to 1-points of $f$. This realization is called a *canonical sum of minterms* realization (which corresponds to the expression for $f$ derived by $n$ iterations of the Shannon expansion theorem).

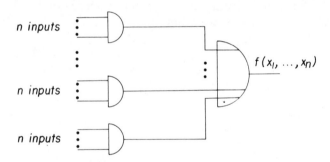

Figure 3.5  Canonical sum of minterms circuit

The function $f_1$ of Figure 3.1 can be expressed as the sum of minterms

$$f_1 = \bar{x}_1 \bar{x}_2 \bar{x}_3 + \bar{x}_1 x_2 \bar{x}_3 + x_1 \bar{x}_2 \bar{x}_3 + x_1 \bar{x}_2 x_3$$
$$= m_0 + m_2 + m_4 + m_5.$$

Incompletely specified functions can be represented by specifying the minterms for which the function is 1 and those for which it is unspecified separately. Thus for the function $f_2$ of Figure 3.1

$$f_2 = \sum_{\substack{1\text{-pts}}} m_0, m_4, m_5 + \sum_{\substack{\text{don't}\\ \text{cares}}} m_2, m_3.$$

Another canonical realization is the *product of maxterms*. If we represent a function $\bar{f}$ as a sum of minterms we obtain $\bar{f} = \Sigma_{i \in I_0} m_i$ where $I_0$ is the set of integer labels corresponding to 0-points of $f$ (or 1-points of $\bar{f}$). Applying DeMorgan's law to the sum of minterms expression for $\bar{f}$, we obtain an expression for $f$ as

$$f = \prod_{i \in I_0} M_i$$

where $M_i = \bar{m}_i$ is a sum term containing each variable complemented or uncomplemented. $M_i$ represents a function which has the value 0 only for the point $m_i$, and hence is called a *maxterm*. The circuit of Figure 3.6 can be used to realize a canonical product of maxterms.

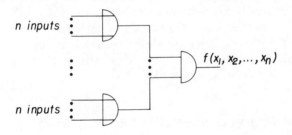

Figure 3.6    Canonical product of maxterms circuit

**Example 3.3**
For the function $f_1$ of Figure 3.1, $\bar{f}_1$ can be expressed as a sum of minterms

$$\bar{f}_1 = m_1 + m_3 + m_6 + m_7.$$

By DeMorgan's Law this is converted to a product of maxterms expression for $f$

$$f = M_1 \cdot M_3 \cdot M_6 \cdot M_7$$
$$= (x_1 + x_2 + \bar{x}_3)(x_1 + \bar{x}_2 + \bar{x}_3)(\bar{x}_1 + \bar{x}_2 + x_3)(\bar{x}_1 + \bar{x}_2 + \bar{x}_3)$$

Note that by interpreting complemented variables as 1's and uncomplemented variables as 0's (the opposite convention as for minterms), the maxterm $M_j$ defines a 0, 1 sequence corresponding to the binary number representation of $j$.

The number of *levels* of a combinational circuit is the maximum number of gates from any input signal to any output. These two canonical realizations are two level circuits. This assumes that each of the variables $x_i$ and its complement $\bar{x}_i$ can be used as circuit inputs. Such inputs are sometimes referred to as *double rail inputs*. (If only the uncomplemented variables can be used as circuit inputs, then inverters may be required and the number of levels in the resultant circuit may exceed two. Such inputs are called *single rail*.)

Since there is usually a delay associated with each gate, the speed of a circuit (i.e. the elapsed time for an input change to cause an output change to occur) is proportional to the number of levels. Hence two level circuits have the desirable property of being relatively fast. However in general there exist other two level realizations for a combinational function which require fewer gates than either of the two canonical realizations previously considered.

## 3.2 COMBINATIONAL FUNCTION MINIMIZATION

For the combinational function $f$, specified by the truth table of Figure 3.7, $f$ can be expressed as a sum of minterms, $f = \bar{x}_1 \bar{x}_2 x_3 + x_1 \bar{x}_2 x_3 + x_1 x_2 x_3$. Using the laws of Boolean algebra, this expression can be simplified since $\bar{x}_1 \bar{x}_2 x_3 + x_1 \bar{x}_2 x_3 + x_1 x_2 x_3 = (\bar{x}_1 + x_1)\bar{x}_2 x_3 + x_1 x_2 x_3$ (by postulate 4(b)) and since $(\bar{x}_1 + x_1) = 1$, this expression simplifies to $\bar{x}_2 x_3 + x_1 x_2 x_3$. Thus two of the three minterms ($m_1$ and $m_5$) can be replaced by a single product term in an equivalent Boolean expression. The original expression could also be simplified by combining

| $x_1$ | $x_2$ | $x_3$ | $f$ |
|-------|-------|-------|-----|
| 0 | 0 | 0 | 0 |
| 0 | 0 | 1 | 1 |
| 0 | 1 | 0 | 0 |
| 0 | 1 | 1 | 0 |
| 1 | 0 | 0 | 0 |
| 1 | 0 | 1 | 1 |
| 1 | 1 | 0 | 0 |
| 1 | 1 | 1 | 1 |

Figure 3.7    A combinational function truth table

minterms $m_5$ and $m_7$ as follows:

$$\bar{x}_1\, \bar{x}_2\, x_3 + x_1\, \bar{x}_2\, x_3 + x_1\, x_2\, x_3 = \bar{x}_1\, \bar{x}_2\, x_3 + x_1\, x_3\, (\bar{x}_2 + x_2)$$

$$= \bar{x}_1\, \bar{x}_2\, x_3 + x_1\, x_3.$$

Both of these simplifications can be utilized if we represent the original expression as $m_1 + m_5 + m_5 + m_7$ ($m_5$ can be repeated since $m_5 + m_5 = m_5$) resulting in the equivalent expression $\bar{x}_2\, x_3 + x_1\, x_3$.

In this section we will consider the problem of deriving two level realizations of a combinational function which are minimal in some respect. There are several criteria of optimality which could be used. The relevance of these criteria is of course dependent on costs associated with the technology which will be used to design the associated circuit. In this chapter we will consider the following two criteria of optimality

(1) A two level circuit which requires the minimal total number of gate inputs.
(2) A two level circuit which requires the minimal number of gates, and has the fewest gate inputs amongst all minimal gate circuits.

In general there may be more than one minimal circuit and therefore we consider the design of "a minimal" rather than "the minimal" circuit. The procedures we will utilize may be extended to derive realizations which are optimal with respect to related criteria of optimality.

We first consider two level realizations referred to as *sum of product* realizations in which the first level consists of AND gates and the second level consists of a single OR gate as shown in Figure 3.8. In such a circuit the AND gates generate products of *literals* where a *literal* is defined to be an input variable $x_i$ or its complement $\bar{x}_i$. The OR gate sums these product terms. Thus the output of this circuit is a sum of product terms.

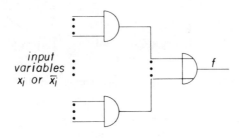

input
variables
$x_i$ or $\bar{x}_i$

$f$

**Figure 3.8    Sum of products circuit**

In an $n$-dimensional space defined by variables $x_1, x_2, \ldots, x_n$, a product term in which each variable or its complement appears, represents a single point in the space as we mentioned previously. Such a term is referred to as a *complete product term*. A product term in which some variable does not appear, either complemented or uncomplemented, represents several points in the $n$-dimensional cube in the following sense. Suppose $P$ is a product term which contains each variable except $x_n$. Then by the law of distributivity, $P = P(x_n + \bar{x}_n) = Px_n + P\bar{x}_n$. Thus $P$ can be expressed as the sum of two complete product terms. Hence for either of these two points in the $n$-cube, $P$ will take the value 1 and hence $P$ is said to *cover* (i.e. represent) these two points. Thus for the set of variables $x_1, x_2, x_3$ the product term $\bar{x}_1 x_3$ covers the two points $(0,0,1)$ $(x_1 = 0, x_2 = 0, x_3 = 1)$ and $(0,1,1)$.

In an exactly analagous manner a product term of $k$ literals covers $2^{n-k}$ points in the $n$-dimensional space. (Such a product term is said to represent a cube of dimension $(n - k)$ or simply an $(n - k)$-cube. Thus a minterm represents a *cube* of dimension 0. For the set of variables $x_1, x_2, x_3$, the product term $\bar{x}_2$ (a product of one literal) covers 4 points, $(0,0,0)$, $(0,0,1)$, $(1,0,0)$, $(1,0,1)$.

For a combinational function $f(x_1, x_2, \ldots, x_n)$ a product term $P$ is an *implicant of f* if $P$ covers at least one 1-point of $f$ and $P$ does not cover any 0-points of $f$.* Thus if $P$ is an implicant of $f$, for every point covered by $P$, $f$ has the value 1 or $f$, is unspecified. An implicant $P$ of a combinational function $f$ is a *prime implicant* of $f$ if for any other implicant of $f$, $P'$, there exists a point which is covered by $P$ but not covered by $P'$.

## Example 3.4

Consider the combinational function $f(x_1, x_2, x_3)$ specified by the truth table of Figure 3.9. The product term $P_1 = \bar{x}_1 \bar{x}_2$ covers the points (0,0,0) and (0,0,1). Since $f$ has the value 1 for both of these points, $P_1$ is an implicant of $f$. Similarly $P_2 = x_1 \bar{x}_2$ covers the points (1,0,0) and (1,0,1). Since $f(1,0,0)$ is unspecified and $f(1,0,1) = 1$, $P_2$ is also an implicant of $f$. The product term $P_3 = \bar{x}_1 x_2$ covers points (0,1,0)

| | $x_1$ | $x_2$ | $x_3$ | $f$ | |
|---|---|---|---|---|---|
| | 0 | 0 | 0 | 1 | $\bar{x}_1 \bar{x}_2 \bar{x}_3 \Rightarrow \bar{x}_1 \bar{x}_2$ |
| | 0 | 0 | 1 | 1 | $x_1 x_2 x \Rightarrow$ |
| | 0 | 1 | 0 | 0 | $x_1 \bar{x}_2 \Rightarrow x_1 \bar{x}_2$ |
| | 0 | 1 | 1 | 1 | |
| $x_1 \bar{x}_2$ | 1 | 0 | 0 | - | |
| | 1 | 0 | 1 | 1 | |
| | 1 | 1 | 0 | 0 | |
| | 1 | 1 | 1 | 0 | |

**Figure 3.9   Combinational function truth table**

and (0,1,1). Since $f(0,1,0) = 0$, $P_3$ is not an implicant of $f$. The product term $P_4 = \bar{x}_2$ covers 4 points and is an implicant of $f$. Since $P_4$ is an implicant and $P_1$ does not cover any point not covered by $P_4$, $P_1$ is not a prime implicant of $f$. Similarly $P_2$ is not a prime implicant of $f$. However $P_4$ is a prime implicant of $f$ since it is the only implicant of $f$ covering 4 points.

The importance of implicants in a sum of products realization of a combinational function $f$ is easily seen by referring to the prototype

---

*Product terms covering only unspecified points have also been defined as implicants in the literature but such terms are superfluous in any sum of products realization.

circuit of Figure 3.8. This circuit will have an output $f = 1$ for any point covered by any product term realized by an AND gate. Hence each such product term must be an implicant of $f$. Since $f = 0$ for any point not covered by some product term, each 1-point of $f$ must be covered by some such product term. Hence in a sum of products realization of $f$ each product term must be an implicant of $f$ and each 1-point of $f$ must be covered by some such product term. The concept of prime implicant is important in sum of products realizations which are optimal with respect to total number of gate inputs or total gates. In such a circuit the output of every AND gate must be a prime implicant of the function being realized as we shall now prove.

**Theorem 3.1**

In a sum of products realization of a combinational function $f$, if the realization is optimal with respect to total gate inputs, or total gates, then the output of every AND gate is a product term which is a prime implicant of $f$.

**Proof:**

Suppose there exists such an optimal realization of $f$ in which the output of some AND gate $G$ is a product term $P_1$ which is not a prime implicant of $f$. Then there must be another implicant of $f$, $P_2$, which covers all points covered by $P_1$. If we replace the AND gate $G$ realizing $P_1$ by an AND gate $G'$ realizing $P_2$, the circuit will still realize $f$ since every point covered by $P_2$ is either unspecified or a 1-point of $f$. If $P_1$ is a product of $k$ literals then $P_2$ is a product of $k' < k$ literals since $P_2$ covers more points than $P_1$. Thus this circuit modification reduces the total number of gate inputs with the same number of gates. This contradicts the assumption that the original circuit was optimal in this respect and hence proves the theorem.

Thus a sum of products realization of a combinational function $f$ which is optimal with respect to gate inputs or gates, is a sum of prime implicants in which each 1-point of $f$ is covered by some prime implicant of the sum. However this optimal realization is not in general equal to the sum of *all* prime implicants of $f$. For instance for the function specified by the truth table of Figure 3.10(a), the set of all prime implicants is $x_1 x_2, \bar{x}_1 x_3, x_2 x_3$. The circuit of Figure 3.10(b) corresponds to the sum of all prime implicants of $f$. However the prime implicants $x_1 x_2$ and $\bar{x}_1 x_3$ cover all 1-points of $f$. Thus $f = x_1 x_2 + \bar{x}_1 x_3$ and the corre-

| $x_1$ | $x_2$ | $x_3$ | $f$ |
|:---:|:---:|:---:|:---:|
| 0 | 0 | 0 | 0 |
| 0 | 0 | 1 | 1 |
| 0 | 1 | 0 | 0 |
| 0 | 1 | 1 | 1 |
| 1 | 0 | 0 | 0 |
| 1 | 0 | 1 | 0 |
| 1 | 1 | 0 | 1 |
| 1 | 1 | 1 | 1 |

(a)

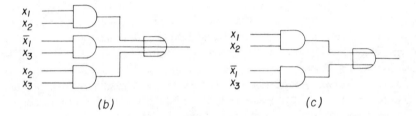

(b)                                                    (c)

**Figure 3.10** (a) Truth table (b) sum of all prime implicants (c) optimal circuit realization

sponding circuit (Figure 3.10(c)) has fewer gate inputs (and fewer gates) than the sum of all prime implicants. From the preceding we can formulate a general procedure for obtaining the minimal gate-input (or minimal gate) sum of products realization of a combinational function $f$.

**Procedure 3.1**
1. Find the complete set of prime implicants of $f$.
2. From the set of all prime implicants of $f$, select a "minimal cost" subset which covers all 1-points of $f$, and realize $f$ as a sum of this minimal cost subset of prime implicants of $f$.

The "cost" of a prime implicant $P_k$ is, of course, related to the optimization goal. A prime implicant of $l$ literals requires an $l$-input AND gate (if $l > 1$) and one input to the OR gate. If $l = 1$ no AND gate is required. Therefore the cost $C(P_k)$ of such a prime implicant is equal to $l + 1$ (if $l > 1$) and the cost is 1 if $l = 1$ assuming the optimization goal is total gate-input minimization. If the optimization goal is to minimize gates, the cost of a prime implicant is 1 if $l >$ 1 and the cost is 0 if $l = 1$. The cost of a set of prime implicants is equal to the sum of the individual costs.

We have now formulated the problem of minimal sum of products combinational circuit synthesis in terms of two well defined subproblems. We shall now consider solutions to these two subproblems.

### 3.2.1 GENERATION OF PRIME IMPLICANTS—TABULAR METHOD

A product term covering two points (i.e. a 1-cube) can be formed by combining two product terms covering one point each (i.e. two 0-cubes) if these two terms differ in only one literal. Suppose these two terms are $x_1 P$ and $\bar{x}_1 P$ where $P$ is a product of literals. Then since $x_1 P + \bar{x}_1 P = (x_1 + \bar{x}_1) P = P$ these two terms are combined by deleting the literal $x_1$ and $\bar{x}_1$ (in which the two terms differ), thus resulting in a product term of $n$-1 literals which covers both points. In a completely analogous manner two product terms of $n$-$k$ literals (each covering $2^k$ points) can be combined if they contain the same set of variables and differ in only one literal, into a composite product term covering $2^{k+1}$ points and having ($n$-$k$-1) literals. Thus $x_2 x_3 \bar{x}_4$ and $x_2 \bar{x}_3 \bar{x}_4$ are combined by deleting $x_3$ and $\bar{x}_3$ into the product term $x_2 \bar{x}_4$. However $x_1 x_2 x_3$ and $x_2 x_3 x_4$ cannot be combined in this manner since the product terms are defined on different sets of variables.

The prime implicants of a combinational function $f$ can be obtained by combining sets of 1-points and unspecified points in this manner. The procedure is iterated until no further combinations are possible at which time the terms which could not be further combined are the complete set of prime implicants of $f$. This procedure can be carried out in a systematic manner as follows.

**Procedure 3.2 (Quine-McCluskey procedure [7,11])**
    (1) List all 1-points and don't care points of $f(x_1, x_2, ..., x_n)$ in a table, representing a minterm $m_i$ by the binary number $i$ arranged as a sequence of 0 and 1 bits in columns labeled $x_1, x_2, ..., x_n$. Partition this list into classes $S_0, S_1, S_2, ..., S_n$ where $S_i$ consists of all such points with $i$ variables equal to 1, and ($n$-$i$) variables equal to 0 in the respective binary representations.

(2) A point in a set $S_i$ and a point in a set $S_j$ where $j > i + 1$ will differ in at least two variables and hence cannot be combined. For all $i = 0,1,2, \ldots,n\text{-}1$, compare each element in set $S_i$ with each element in set $S_{i+1}$. For those pairs of points which differ in a single literal $x_j$, create a new implicant which agrees with both points in the $(n\text{-}1)$ literals in which they agree and is unspecified $(-)$ in $x_j$ and place it in a new set $S_i'$ containing 1-cubes with $i$ variables equal to 1. To indicate that the two individual points which were combined are not prime implicants, flag $(\checkmark)$ each point when it is combined with another point. Repeat this for all pairs of points of which one is in $S_i$ and the other in $S_{i+1}$ for all $i = 0,1,2, \ldots,n\text{-}1$.

(3) After completing (2) apply the same procedure to the sets $S_0', S_1', \ldots, S_{n-1}'$, to combine implicants. Note that implicants with don't cares must have the don't cares in the same variables in order to be combined. Implicants are flagged $(\checkmark)$ when combined with other implicants). This results in sets $S_0'', S_1'', \ldots, S_{n-2}''$. The procedure is then iterated on these sets continually until no further combinations are possible, at which time all unflagged elements represent prime implicants of $f$.*

The representation of minterms as binary numbers enables the determination of the number of bits in which terms differ by inspection. Alternatively it is possible to represent minterms as decimal numbers. Such terms differ in one variable only if the difference between their decimal representations is $2^k$ for some integer $k$. Implicants representing a set of $2^r$ points can be represented by a set of decimal minterms. Two implicants can be combined only if each element of one set differs from some element of the other set by $2^k$ for the same value of $k$. For functions of many variables, the decimal procedure for determination of prime implicants, may result in fewer errors of a mechanical nature for hand computations.

### Examples 3.5

Consider the combinational function $f(x_1,x_2,x_3,x_4)$, specified by the truth table of Figure 3.11(a). The 1-points and the don't care point of $f$ are listed in a table and partitioned into sets $S_0,S_1,S_2,S_3,S_4$ as shown

---

*For some of these terms all points covered may be unspecified, and hence these are not actually prime implicants. Such terms will be discarded during step (2) of Procedure 3.1, the selection of a minimal cost set of prime implicants which cover all 1-points of $f$.

76 Combinational Circuit Synthesis

| $x_1$ | $x_2$ | $x_3$ | $x_4$ | $f$ |
|---|---|---|---|---|
| 0 | 0 | 0 | 0 | 1 |
| 0 | 0 | 0 | 1 | 1 |
| 0 | 0 | 1 | 0 | 1 |
| 0 | 0 | 1 | 1 | 0 |
| 0 | 1 | 0 | 0 | 0 |
| 0 | 1 | 0 | 1 | 1 |
| 0 | 1 | 1 | 0 | 0 |
| 0 | 1 | 1 | 1 | - |
| 1 | 0 | 0 | 0 | 1 |
| 1 | 0 | 0 | 1 | 0 |
| 1 | 0 | 1 | 0 | 1 |
| 1 | 0 | 1 | 1 | 0 |
| 1 | 1 | 0 | 0 | 0 |
| 1 | 1 | 0 | 1 | 0 |
| 1 | 1 | 1 | 0 | 1 |
| 1 | 1 | 1 | 1 | 1 |

(a)

|  | $x_1$ | $x_2$ | $x_3$ | $x_4$ |  |
|---|---|---|---|---|---|
| $S_0$ { | 0 | 0 | 0 | 0✓ | $m_0$ |
| $S_1$ { | 0 | 0 | 0 | 1✓ | $m_1$ |
|  | 0 | 0 | 1 | 0✓ | $m_2$ |
|  | 1 | 0 | 0 | 0✓ | $m_8$ |
| $S_2$ { | 0 | 1 | 0 | 1✓ | $m_5$ |
|  | 1 | 0 | 1 | 0✓ | $m_{10}$ |
| $S_3$ { | 0 | 1 | 1 | 1✓ | $m_7$ |
|  | 1 | 1 | 1 | 0✓ | $m_{14}$ |
| $S_4$ | 1 | 1 | 1 | 1✓ | $m_{15}$ |
| $S_0'$ { | 0 | 0 | 0 | - | $(m_0,m_1)\ \Delta = 1$ |
|  | 0 | 0 | - | 0✓ | $(m_0,m_2)\ \Delta = 2$ |
|  | - | 0 | 0 | 0✓ | $(m_0,m_8)\ \Delta = 8$ |
| $S_1'$ { | 0 | - | 0 | 1 | $(m_1,m_5)\ \Delta = 4$ |
|  | - | 0 | 1 | 0✓ | $(m_2,m_{10})\ \Delta = 8$ |
|  | 1 | 0 | - | 0✓ | $(m_8,m_{10})\ \Delta = 2$ |
| $S_2'$ { | 0 | 1 | - | 1 | $(m_5,m_7)\ \Delta = 2$ |
|  | 1 | - | 1 | 0 | $(m_{10},m_{14})\ \Delta = 4$ |
| $S_3'$ { | - | 1 | 1 | 1 | $(m_7,m_{15})\ \Delta = 8$ |
|  | 1 | 1 | 1 | - | $(m_{14},m_{15})\ \Delta = 1$ |
| $S_0''$ { | - | 0 | - | 0 | $(m_0,m_2,m_8,m_{10})\ \Delta_1 = 2,\ \Delta_2 = 8$ |

(b)                                (c)

Figure 3.11   (a) Truth table (b) binary prime implicant generation (c) decimal prime implicant generation in Figure 3.11(b). Set $S_0'$ is formed by combining elements in $S_0$ and $S_1$. The first row in $S_0'$ results from combining the points (0000) and (0001), the second row from combining (0000) and (0010), and the third from (0000) and (1000). Note that the same row is combined with several other rows. Similarly $S_1'$ is formed from $S_1$ and $S_2$, $S_2'$ from $S_2$ and $S_3$, and $S_3'$ from $S_3$ and $S_4$. The 3rd row of $S_0'$ and the 2nd row of $S_1'$ are combined to form $S_0''$. Combining row 2 of $S_0'$ and row 3 of $S_1'$ results in the same implicant. The unflagged rows are the prime implicants of $f$: $\bar{x}_1\,\bar{x}_2\,\bar{x}_3\,,\bar{x}_1\,\bar{x}_3\,x_4\,,\bar{x}_1\,x_2\,x_4\,,x_1\,x_3\,\bar{x}_4\,,x_2\,x_3\,x_4\,,x_1\,x_2\,x_3\,,\bar{x}_2\,\bar{x}_4$.

Figure 3.11(c) shows the procedure for determining prime implicants, using a decimal notation. Associated with each row of each set $S_i'$ and $S_i''$, etc. is the set of points covered by the corresponding implicant and the decimal difference $\Delta$ between these elements. The rows $(m_0,m_2)$ of $S_0'$ and $(m_8,m_{10})$ of $S_1'$ have constant difference $\Delta = 8 = 2^3$ and are combined to form a set of 4 elements $(m_0,m_2,m_8,m_{10})$ with 2 differences $\Delta_1 = 2$ (the difference for the original combinations $(m_0,m_2)$ and $(m_8,m_{10})$ and $\Delta_2 = 8$ (the constant difference between these 2 sets). A row in $S_i'$ can be combined with a row of $S_{i+1}'$ if and only if the difference (or set of differences) of these two rows is identical.

## 3.2.2 GENERATION OF PRIME IMPLICANTS—MAP METHOD

Another method of generating prime implicants for functions of a relatively small number of variables involves the use of a Karnaugh map [4]. A Karnaugh map for $n$ input variables $(x_1,x_2,...,x_n)$ is effectively a 2-dimensional pictorial representation of the $n$-cube represented by the $n$ inputs. Such a map contains $2^n$ cells, one representing each of the $2^n$ points in the $n$-cube (i.e. each of the $2^n$ possible values of the inputs $(x_1,x_2,...,x_n)$). This map is defined in such a way that orthogonal adjacent cells in the map differ in only one variable. Karnaugh maps for functions of three, four and five variables are shown in Figure 3.12. Note that there are several different maps possible for each of these. For example, two different four-variable maps are shown in Figure 3.12(b) and (c). The input combination represented by each cell of the map can be determined by examining the column and row labels. For example, the point labeled $a$ in the maps of Figure 3.12(b) and (c) represents the input combination 0010, since it is not within either of the two columns labeled $x_1 = 1$, not in either of the two columns labeled $x_2 = 1$, or in the two rows labeled $x_4 = 1$, but is in a row labeled $x_3$

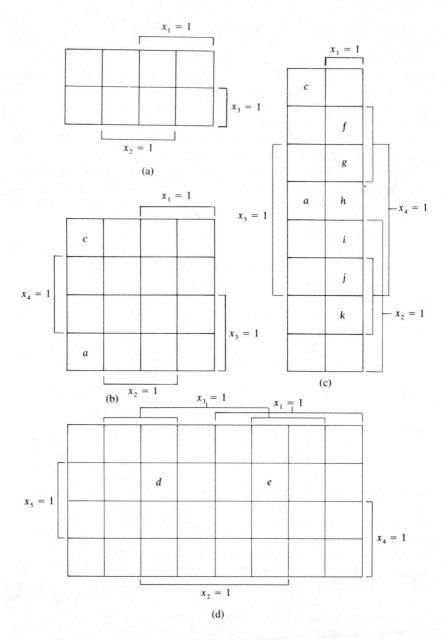

Figure 3.12   Karnaugh maps of (a) 3 variables, (b) 4 variables, (c) 4 variables, (d) 5 variables

= 1. Note that exactly one half of the cells in each map correspond to $x_i = 1$ and the remaining cells to $x_i = 0$ for each of the $n$ variables $x_1, x_2, \ldots, x_n$.

For each point $q$ in an $n$-dimensional space ($n$-cube) there are exactly $n$ points which differ from $q$ in only one variable. From the maps of Figure 3.12 it is obvious that each cell can have at most four orthogonal neighbor cells. Thus it is necessary to use non-adjacent cells in the map to represent points that differ in only one variable. In order for the maps to be useful for determining prime implicants, it should be possible to determine points that differ in a single variable by inspection. For example, the points $a$ and $c$ in Figure 3.12(b) (and (c)) differ only in the variable $x_3$. In this map, the first and fourth columns may be treated as being adjacent and likewise, the top and bottom rows. The usefulness of Karnaugh maps is limited to a small number of variables, because of our inability to represent larger cubes in such a way that neighboring input combinations can be determined by inspection.

A combinational function can be represented on a Karnaugh map by filling in the 1 and don't care entries of the map with the understanding that the function has the value 0 at the other points. The function specified by the table of Figure 3.11(a) (Example 3.5) can thus be represented by the Karnaugh map of Figure 3.13.

Since implicants of a function are formed by combining 1-points of a function which differ in only one input variable, such terms will appear

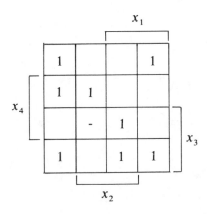

Figure 3.13  Karnaugh map of a combinational function

in a Karnaugh map as a cluster of "adjacent" 1-points, where points such as $a$ and $c$ in Figure 3.12(b) are considered to be adjacent. With a little experience, it becomes easy to visually identify such clusters of $2^k$ adjacent points which represent implicants. In general, $k$-cubes can be determined by their symmetry with respect to $k$ axes in the map. Thus $a$ and $c$ in Figure 3.12(b) are symmetric with respect to the axis separating the $x_3 = 1$ and $x_3 = 0$ spaces, and $d$ and $e$ in Figure 3.12(d) are symmetric with respect to the axis separating the $x_1 = 1$ and $x_1 = 0$ spaces. In Figure 3.12(c) the set of points $\{f,g,j,k\}$ forms a cube of four points (a 2-cube) that is symmetric with respect to the $x_2$ and $x_3$ axes. Since $x_1 = 1$ and $x_4 = 1$ for all four points, they represent the implicant $x_1 x_4$.

The prime implicants of a function can be determined by finding all $k$-cubes that are not contained in larger cubes

**Example 3.6**

The combinational function specified in the truth table of Figure 3.11(a) (Example 3.5) is represented by the Karnaugh map of Figure 3.13. In Figure 3.14(a) and (b), clusters of 1-points defining prime implicants are encircled and labelled. These prime implicants are

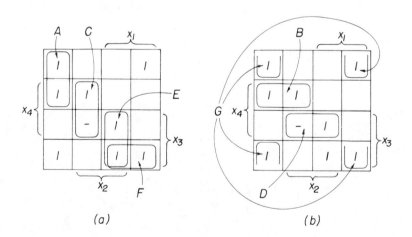

(a)                              (b)

Figure 3.14   Karnaugh map of prime implicants of a function

$$A: \bar{x}_1 \bar{x}_2 \bar{x}_3$$

$$B: \bar{x}_1 \bar{x}_3 x_4$$

$$C: \bar{x}_1 x_2 x_4$$

$$D: x_2 x_3 x_4$$

$$E: x_1 x_2 x_3$$

$$F: x_1 x_3 \bar{x}_4$$

$$G: \bar{x}_2 \bar{x}_4$$

which is identical to the set of prime implicants obtained in Example 3.5 using the tabular procedure. The prime implicant $\bar{x}_2 \bar{x}_4$ is defined by the four corner cells.

### 3.2.3 SELECTION OF A MINIMAL COVERING SET OF PRIME IMPLICANTS

The problem of selecting a minimal cost set of prime implicants which cover all 1-points of a function $f$, is referred to as the *prime implicant covering problem*. The covering relationship between prime implicants and 1-points can be represented by a *prime implicant covering table*, in which there is one row corresponding to each prime implicant, one column corresponding to each minterm which is a 1-point of $f$, and the entry in row $i$ and column $j$ is 1 if and only if the prime implicant represented by row $i$ covers the 1-point represented by column $j$. Associated with each row (prime implicant) is a cost, which reflects the cost of the hardware associated with realizing that prime implicant as part of a sum of products circuit. Cost functions for minimal gate inputs and for minimal gates as optimality objectives have been previously discussed. These or other cost functions can be associated with the rows of a covering table. For the combinational function of Example 3.6, the prime implicant covering table is shown in Figure 3.15, where column $C_1$ reflects the cost function associated with minimization of gate inputs and $C_2$, the cost function associated with minimization of gates. Note that the unspecified entry of $f$, $m_7$, does not define a column of the covering table since it is not necessary to cover that point.

The *prime implicant covering problem* is to select a minimal cost set of rows (prime implicants) such that for every column (1-point) $m_j$,

| | $m_0$ | $m_1$ | $m_2$ | $m_5$ | $m_8$ | $m_{10}$ | $m_{14}$ | $m_{15}$ | $C_1$ | $C_2$ |
|---|---|---|---|---|---|---|---|---|---|---|
| $(P_1)\ \bar{x}_1\,\bar{x}_2\,\bar{x}_3$ | 1 | 1 | | | | | | | 4 | 1 |
| $(P_2)\ \bar{x}_1\,\bar{x}_3\,x_4$ | | 1 | | 1 | | | | | 4 | 1 |
| $(P_3)\ \bar{x}_1\,x_2\,x_4$ | | | 1 | | | | | | 4 | 1 |
| $(P_4)\ x_1\,x_3\,\bar{x}_4$ | | | | | 1 | 1 | | | 4 | 1 |
| $(P_5)\ x_1\,x_2\,x_3$ | | | | | | | 1 | 1 | 4 | 1 |
| $(P_6)\ x_2\,x_3\,x_4$ | | | | | | | | 1 | 4 | 1 |
| $(P_7)\ \bar{x}_2\,\bar{x}_4$ | 1 | | 1 | | 1 | 1 | | | 3 | 1 |

Figure 3.15   Prime implicant covering table

at least one row in the selected set has a 1 in column $m_j$ (i.e. covers $m_j$), where the cost of a set of rows is the arithmetic sum of the costs of the individual rows in the set.

We shall now present some rules which sometimes are helpful in solving the covering problem. These rules may determine some rows which can be selected as part of a minimal cost cover, and may simplify the table via elimination of rows and/or columns in such a way that any minimal cost covering set for the simplified (reduced) table is also a minimal cost covering set for the original table.

### Reduction Rule 1 (Essential Prime Implicant)

A prime implicant $P_i$ of a combinational function $f$ is *essential with respect to a 1-point $m_j$* of $f$ if $P_i$ is the only prime implicant which covers $m_j$. An essential prime implicant corresponds to a row $P_i$ of the covering table such that the only 1-entry in column $m_j$ occurs in row $P_i$. Any cover of the table must contain $P_i$. Therefore, row $P_i$ can be selected as a member of the minimal cost covering set. The table can then be simplified by deleting row $P_i$ and all columns covered by $P_i$.

For the table of Figure 3.15, $P_7$ is essential with respect to $m_8$ (and $m_2$). This row can be selected as part of the minimal covering set and along with columns $m_0, m_2, m_8, m_{10}$ can be deleted from the table. The reduced table is shown in Figure 3.16 This table cannot be further simplified by Reduction Rule 1.

|       | $m_1$ | $m_5$ | $m_{14}$ | $m_{15}$ | $C_1$ | $C_2$ |
|-------|-------|-------|----------|----------|-------|-------|
| $P_1$ | 1     |       |          |          | 4     | 1     |
| $P_2$ | 1     | 1     |          |          | 4     | 1     |
| $P_3$ |       | 1     |          |          | 4     | 1     |
| $P_4$ |       |       | 1        |          | 4     | 1     |
| $P_5$ |       |       | 1        | 1        | 4     | 1     |
| $P_6$ |       |       |          | 1        | 4     | 1     |

Figure 3.16    Reduced prime implicant covering table

**Reduction Rule 2 (Row Dominance)**

Let $C(P_k)$ be the cost of row (prime implicant) $P_k$. A row $P_i$ *dominates* a row $P_j$ if $C(P_i) \leq C(P_j)$ *and* $P_i$ covers all 1-points covered by $P_j$ (i.e., $P_i$ has 1's in all columns in which $P_j$ has 1's). If row $P_i$ dominates $P_j$, there exists a minimal cost cover which does not contain $P_j$ (Exercise). Therefore, we can eliminate $P_j$ from the table since any minimal cost cover for the reduced table will also be a minimal cost cover for the original table.

The table reduction resulting from application of this rule may cause a row of the table to become essential. Rule 1 can then be applied to further simplify the table. For the table of Figure 3.16, for either cost function, row $P_5$ dominates both $P_4$ and $P_6$ and row $P_2$ dominates both $P_1$ and $P_3$. Thus rows $P_1$, $P_3$, $P_4$ and $P_6$ can be eliminated from the table, resulting in the reduced table of Figure 3.17. In this table $P_2$ is essential with respect to $m_1$ and $m_5$ and $P_5$ is essential with respect to $m_{14}$ and $m_{15}$. Hence the minimal covering set for the table of Figure 3.17 consists of $P_2$ and $P_5$ and the minimal covering set for

|       | $m_1$ | $m_5$ | $m_{14}$ | $m_{15}$ | $C_1$ | $C_2$ |
|-------|-------|-------|----------|----------|-------|-------|
| $P_2$ | 1     | 1     |          |          | 4     | 1     |
| $P_5$ |       |       | 1        | 1        | 4     | 1     |

Figure 3.17    Reduced covering table

the table of Figure 3.15 consists of $P_7$, $P_2$ and $P_5$.

These two reduction rules may not be sufficient to determine a minimal covering set for some covering tables, if the original table or a reduced version of the original table has no essential or dominating rows. Such tables are called *cyclic tables*. A procedure called *branching* is useful for obtaining a minimal cover for such tables. The procedure consists of selecting a column $m_j$ of the table and the set of rows $R$ which cover $m_j$. Any covering set must contain one or more of these rows. We arbitrarily choose one row in $R$ as part of the cover, simplify the table accordingly by deleting this row and all columns covered by it, and obtain the optimal cover containing this row. We then repeat this operation for each row in the set $R$. We compare all of these "optimal" covers and select the lowest cost one as the minimal cost cover of the original cyclic table. Thus as the name implies, the branching procedure consists of exhaustive examination of a set of different rows one of which is required to be part of a cover.

Branching can be performed with respect to any column of the cyclic table. However, in general, it seems computationally more efficient to branch around the column with the fewest number of 1-entries. In addition, several levels of branching may be required to obtain a minimal cost cover. That is in obtaining the minimal cover containing a particular row of a set $R$, where $R$ is defined by a column $m_j$, it may be necessary to branch around another set $R'$ defined by another column $m_p$. The following example demonstrates the application of the branching procedure to a cyclic table.

**Example 3.7**

In the cyclic table of Figure 3.18 all rows have cost $k$. From column $m_a$, we can determine that any cover contains $P_1$ and/or $P_3$. We apply the branching procedure and first determine the optimal cover containing $P_1$. Eliminating row $P_1$ and columns $m_a$, $m_b$ and $m_d$, which are covered by $P_1$, results in the reduced table of Figure 3.19. In this table row $P_5$ dominates $P_2$, $P_3$, $P_4$ and $P_6$. Hence the optimal cover containing $P_1$ for the table of Figure 3.18 is $P_1$ and $P_5$ which has cost $2k$.

We now determine the optimal cover containing $P_3$ for the cyclic table of Figure 3.18. In this case $P_3$ is selected and the table is reduced to that of Figure 3.20. For this table $P_1$ dominates $P_2$, and $P_4$ dominates $P_5$. After this reduction the table is again cyclic. An optimal cover containing $P_3$ for the cyclic table of Figure 3.20 consists of $P_1$, $P_3$ and any 2 of the rows $(P_1, P_4, P_6)$ and has cost $3k$.

|       | $m_a$ | $m_b$ | $m_c$ | $m_d$ | $m_e$ | $C$ |
|-------|-------|-------|-------|-------|-------|-----|
| $P_1$ | 1     | 1     |       | 1     |       | $k$ |
| $P_2$ |       | 1     | 1     |       |       | $k$ |
| $P_3$ | 1     |       | 1     |       |       | $k$ |
| $P_4$ |       |       |       | 1     | 1     | $k$ |
| $P_5$ |       |       | 1     |       | 1     | $k$ |
| $P_6$ |       | 1     |       |       | 1     | $k$ |

Figure 3.18   A cyclic table

|       | $m_c$ | $m_e$ | $C$ |
|-------|-------|-------|-----|
| $P_2$ | 1     |       | $k$ |
| $P_3$ | 1     |       | $k$ |
| $P_4$ |       | 1     | $k$ |
| $P_5$ | 1     | 1     | $k$ |
| $P_6$ |       | 1     | $k$ |

Figure 3.19   The reduced table, branch 1

|       | $m_b$ | $m_d$ | $m_e$ | $C$ |
|-------|-------|-------|-------|-----|
| $P_1$ | 1     | 1     |       | $k$ |
| $P_2$ | 1     |       |       | $k$ |
| $P_4$ |       | 1     | 1     | $k$ |
| $P_5$ |       |       | 1     | $k$ |
| $P_6$ | 1     |       | 1     | $k$ |

Figure 3.20   The reduced table, branch 2

Since the other branch generated a cover of cost $2k$, the minimal cost cover for the original table consists of $P_1$ and $P_5$ and is of cost $2k$.

One additional reduction rule for a covering table is that of *column dominance.*

### Reduction Rule 3 (Column Dominance)

A *column $m_i$ of a covering table dominates a column $m_j$ if every row which covers $m_j$ also covers $m_i$* (i.e. column $m_i$ has 1's in all rows in which column $m_j$ has 1's). Any row which covers $m_j$ will also cover $m_i$. Column $m_i$ can thus be eliminated from the table since any covering set of rows for the reduced table will cover $m_j$ and hence will also cover $m_i$.

However any table which is cyclic with respect to the first two rules will also be cyclic with respect to all three rules. Small prime implicant table covering problems can also be solved by a method developed by Petrick[9]. Each column of the table can be covered by any prime implicant which has a 1 in that column. For each prime implicant we define a Boolean variable $p_i$ such that $p_i = 1$ if the prime implicant $P_i$ is a member of the covering set, and $p_i = 0$ otherwise. Then the condition that a column $m_j$ be covered by some prime implicant can be represented by the Boolean expression $\Sigma_{i \in I_j} p_i = 1$, where $I_j$ is the set of rows covering the column $m_j$. For example, column $m_a$ of the table of Figure 3.18 will be covered if $(p_1 + p_3) = 1$. Similarly covering column $m_b$ requires $(p_1 + p_2 + p_6) = 1$. The set of prime implicants covering both columns $m_a$ and $m_b$ must satisfy the condition

$$(p_1 + p_3) \cdot (p_1 + p_2 + p_6) = 1.$$

Since all the $p_i$'s are Boolean variables, this condition can be written as:

$$p_1 + p_2 p_3 + p_3 p_6 = 1.$$

In general, the condition for covering all columns of an $m$-column table can be written as the Boolean expression

$$\prod_{j=1}^{m} \left( \sum_{i \in I_j} p_i \right) = 1.$$

If the left hand side of this equation is written in a sum of products form, each term will represent a set of prime implicants that cover all 1-points and the expression will represent *all covers* which are not subsets of other covers. The cost of each set is the sum of the costs of the individual prime implicants contained in the set. If all prime implicants have equal cost the product term with the fewest number of variables represents a minimal cover. Otherwise a minimal cover is defined by the lowest cost product term.

For the table of Figure 3.18, this procedure generates the expression

$$(p_1 + p_3)(p_1 + p_2 + p_6)(p_2 + p_3 + p_5)(p_1 + p_4)(p_4 + p_5 + p_6)$$
$$= p_1 p_5 + p_1 p_2 p_4 + p_1 p_2 p_6 + p_2 p_3 p_4 + p_3 p_4 p_6 + p_1 p_3 p_4$$
$$+ p_1 p_3 p_6.$$

The minimal cost cover is defined by the product term $p_1 p_5$.

The concepts of essentiality and dominance can frequently be visually applied to derive essential prime implicants and *good* prime implicants directly from a Karnaugh map representation of a function $f$. A prime implicant $P$ is *good* with respect to a 1-point $m_j$ of a combinational function $f$ if $P$ covers $m_j$ and $P$ dominates all other prime implicants which cover $m_j$ (i.e. $P$ covers all (previously uncovered) 1-points covered by any other prime implicant $P'$ which covers $m_j$, and $C(P) \le C(P')$). For any prime implicant $P$ which is good with respect to a 1-point $m_j$ of a function $f$, a minimal sum of products realization can be obtained containing $P$ and hence, $P$ can be selected as part of a minimal realization. Once an essential or good prime implicant has been selected as part of the minimal cover, the 1-points covered by this term are changed to unspecified entries in the Karnaugh map and other good prime implicants can be selected from the resulting map. Thus the Karnaugh map can sometimes be used to combine the two steps of prime implicant generation and selection in the derivation of minimal sum of products realizations. Of course, branching may sometimes be required.

## Example 3.8

For the combinational function $f$ specified by the Karnaugh map of Figure 3.21, the 1-point 0010 is covered by only one prime implicant $\bar{x}_2 x_3 \bar{x}_4$ which is therefore essential with respect to $m_2$ of $f$. The 1-point 0111 is covered by two prime implicants, $\bar{x}_1 x_2 x_4$ and $x_2 x_3 x_4$. However $x_2 x_3 x_4$ does not cover any other 1-points of $f$. Therefore $\bar{x}_1 x_2 x_4$ is good with respect to $m_7$ of $f$ and is selected as a product term in

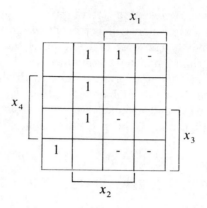

Figure 3.21   A combinational function map

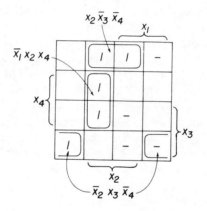

Figure 3.22   Prime implicants of covering set

the minimal gate input sum of products of $f$. (Note that $\bar{x}_1 x_2 x_4$ is not good with respect to the other 1-point it covers, $m_5$). The 1-point 0100 is covered by two prime implicants, $x_2 \bar{x}_3 \bar{x}_4$ and $\bar{x}_1 x_2 \bar{x}_3$ each of which covers two 1-points of $f$. However the other 1-point covered by $\bar{x}_1 x_2 \bar{x}_3$, 0101, has already been covered by the previously selected prime implicant $\bar{x}_1 x_2 x_4$. Therefore $x_2 \bar{x}_3 \bar{x}_4$ is good with respect to $m_4$ of $f$. (Note that $x_2 \bar{x}_3 \bar{x}_4$ is not good with respect to the 1-point 1100 since the prime implicant $x_1 \bar{x}_4$ covers this point and has fewer literals.) At this stage all 1-points have been covered and the minimal cost sum of products expression is $f = \bar{x}_2 x_3 \bar{x}_4 + \bar{x}_1 x_2 x_4 + x_2 \bar{x}_3 \bar{x}_4$. The three prime implicants comprising this covering set are depicted by the circled sets of points in the Karnaugh map of Figure 3.22.

It is conjectured that for completely specified functions, a minimum gate input sum of products realization is also a minimum gate sum of products realization. However this is not always true for incompletely specified functions, as illustrated by the following example.

**Example 3.9**

Consider the function $f(x_1, x_2, ..., x_{10})$, which is defined as follows:

$$f(0,0,0,0,0,0,0,0,0,0) = f(1,1,0,0,0,0,0,0,0,0) = 1$$

$f(0,1,0,0,0,0,0,0,0,0)$ and $f(1,0,0,0,0,0,0,0,0,0)$ are unspecified

$$f(0,1,x_3,x_4, ...,x_{10}) = 0 \text{ if } x_3 + x_4 + ... + x_{10} = 1$$

$$f(1,0,x_3,x_4, ...,x_{10}) = 0 \text{ if } x_3 + x_4 + ... + x_{10} = 1$$

For all other points $f$ is unspecified. This function has 3 prime implicants. $P_1 = \bar{x}_1 \bar{x}_2$, $P_2 = x_1 x_2$, $P_3 = \bar{x}_3 \bar{x}_4 \bar{x}_5 \bar{x}_6 \bar{x}_7 \bar{x}_8 \bar{x}_9 \bar{x}_{10}$ .

The covering table is as shown in Figure 3.23, where column $C_1$ is the cost function to minimize gate inputs and $C_2$ is the cost function to minimize gates. The minimal cost gate-input realization is $f = \bar{x}_1 \bar{x}_2 + x_1 x_2$, defined by the covering set $\{P_1, P_2\}$. The minimal cost gate realization is $f = \bar{x}_3 \bar{x}_4 \bar{x}_5 \bar{x}_6 \bar{x}_7 \bar{x}_8 \bar{x}_9 \bar{x}_{10}$ which is defined by the covering set $\{P_3\}$.

|       | $m_0$ | $m_2$ | $C_1$ | $C_2$ |
|-------|-------|-------|-------|-------|
| $P_1$ | 1     |       | 3     | 1     |
| $P_2$ |       | 1     | 3     | 1     |
| $P_3$ | 1     | 1     | 8     | 1     |

Figure 3.23   The covering table of Example 3.9

## 3.2.4 MINIMAL PRODUCT OF SUMS REALIZATIONS

We have thus shown how a minimal sum of products realization can be derived for a function $f$. A product of sums realization (OR/AND) of $f$ can be derived from a sum of products realization of $\bar{f}$ by DeMorgan's Law as follows

If
$$\bar{f} = \sum_{i \in I} P_i \qquad \text{then} \qquad f = \prod_{i \in I} \bar{P}_i$$

where if $P_i = \Pi x_k^*$ then $\bar{P}_i = \Sigma \bar{x}_k^*$ (where $x_k^*$ is $x_k$ or $\bar{x}_k$). (If $P_i$ is a prime implicant of $\bar{f}$, then $\bar{P}_i$ is called a *prime implicate* of $f$.)

If this transformation is applied to a *minimal cost* sum of products realization of $\bar{f}$, the resulting circuit is a minimal cost product of sums realization of $f$ (for either of the two cost functions being considered).

## Theorem 3.2

The expression derived by applying DeMorgan's law to a minimal sum of products realization of a function $\bar{f}$ is a minimal product of sums realization of $f$.

## Proof

Suppose there is a product of sums expression for $f$, $E'$, with cost $C'$ and the cost of the product of sums expression $E$ derived from $G$, the minimal sum of products expression of $\bar{f}$, is $C > C'$. Then applying DeMorgan's law to $E'$ we derive a sum of products expression of $\bar{f}$ with cost $C'$ (Exercise). This contradicts the assumption that $G$ is a minimal sum of products expression of $\bar{f}$ and proves the theorem.

Hence the following procedure can be used to generate a minimal product of sums realization for a combinational function $f$.

## Procedure 3.3

1) Define the combinational function $\bar{f}$ as follows:
For a point $m_i$ in the $n$-cube

if $\qquad f(m_i) = 1$ then $\bar{f}(m_i) = 0$,

if $\qquad f(m_i) = 0$ then $\bar{f}(m_i) = 1$,

if $\qquad f(m_i)$ is unspecified then $\bar{f}(m_i)$ is unspecified.

2) Using Procedure 3.1 find a minimal cost sum of products realization of $\bar{f}$, $\bar{f} = \Sigma_{i \in I} P_i$.

3) Apply DeMorgan's Law to this expression to obtain $f = \Pi_{i \in I} \bar{P}_i$.

The resulting expression is a minimal cost product of sums realization of $f$.

In general the minimal cost sum of products realization of a function $f$, and the minimal cost product of sums realization of $f$ will not have equal costs. Thus to determine a minimal cost two level realization of $f$ we would determine both the minimal cost sum of products and the minimal cost product of sums realizations. The two level realization of minimal cost is the less costly of these two realizations.

**Example 3.10**

For the combinational function specified by the Karnaugh map of Figure 3.24, the minimal gate input sum of products realization is obtained as follows:

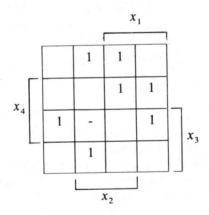

**Figure 3.24   A combinational function map**

The 1-point 0110 is covered by two prime implicants, $\bar{x}_1 x_2 x_3$ and $\bar{x}_1 x_2 \bar{x}_4$. Since $\bar{x}_1 x_2 x_3$ does not cover any other 1-point of $f$, $\bar{x}_1 x_2 \bar{x}_4$ is good with respect to $m_6$ of $f$. The 1-point 1100 is covered by $x_1 x_2 \bar{x}_3$ and $x_2 \bar{x}_3 \bar{x}_4$. However $x_2 \bar{x}_3 \bar{x}_4$ does not cover any other 1-point except 0100 and that was previously covered by $\bar{x}_1 x_2 \bar{x}_4$. Therefore $x_1 x_2 \bar{x}_3$ is good with respect to $m_{12}$ of $f$. Similarly $x_1 \bar{x}_2 x_4$ is then good with respect to $m_9$ and $\bar{x}_2 x_3 x_4$ is good with respect to $m_3$. Thus a minimal sum of products expression for $f$ is $f = \bar{x}_1 x_2 \bar{x}_4 + x_1 x_2 \bar{x}_3 + x_1 \bar{x}_2 x_4 + \bar{x}_2 x_3 x_4$.

The corresponding realization has four 3-input AND gates and one 4-input OR gate, thus requiring 16 gate inputs.

The function $\bar{f}$ is shown in the Karnaugh map of Figure 3.25. The minimal sum of products of $\bar{f}$ is derived in order to determine the minimal product of sums of $f$. In the map of Figure 3.25, the only prime implicant

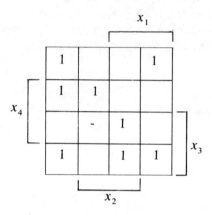

**Figure 3.25   The map of $\bar{f}$**

which covers the 1-point 1000 is $\bar{x}_2 \, \bar{x}_4$. Therefore $\bar{x}_2 \, \bar{x}_4$ is essential with respect to $m_8$. (It is also essential with respect to $m_2$ but not with respect to $m_0$ or $m_{10}$). The 1-point 1111 is covered by two prime implicants, $x_1 x_2 x_3$ and $x_2 x_3 x_4$. However since $x_2 x_3 x_4$ does not cover any other 1-points, $x_1 x_2 x_3$ is good with respect to $m_{15}$. Similarly $\bar{x}_1 \bar{x}_3 x_4$ is good with respect to $m_5$. The minimal sum of products expression for $\bar{f}$ is $\bar{f} = \bar{x}_2 \bar{x}_4 + x_1 x_2 x_3 + \bar{x}_1 \bar{x}_3 x_4$. By applying DeMorgan's Law to this expression we obtain the minimal product of sums expression, $f = (x_2 + x_4)(\bar{x}_1 + \bar{x}_2 + \bar{x}_3)(x_1 + x_3 + \bar{x}_4)$. This requires one 2-input OR gate, two 3-input OR gates and one 3-input AND gate, 11 gate inputs in all. Hence for this function, the minimal cost two level realization is the minimal cost product of sums.

The minimum gate-input two level realization may not be a minimum gate two level realization even for completely specified functions as illustrated in the following example.

**Example 3.11**

For the function $f$ of the Karnaugh map of Figure 3.26, the minimum gate input two level realization is the product of sums $f = (\bar{x}_1 + x_2)(\bar{x}_1 + x_3 + \bar{x}_4)(\bar{x}_3 + x_4)(x_2 + \bar{x}_4)(x_1 + \bar{x}_3)$ which requires 6 gates and 16 gate inputs. The minimal sum of products realization is $f = x_1 x_2 x_3 x_4$

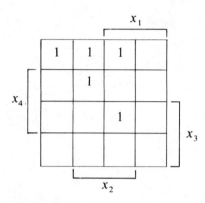

Figure 3.26    The function of Example 3.11

$+ \bar{x}_1 x_2 \bar{x}_3 + x_2 \bar{x}_3 \bar{x}_4 + \bar{x}_1 \bar{x}_3 \bar{x}_4$ which requires 5 gates and 17 gate inputs. Hence the sum of products realization is minimal with respect to number of gates while the product of sums is minimal with respect to number of gate inputs.

## 3.2.5 MINIMAL TWO LEVEL *NAND* AND *NOR* REALIZATIONS

It is also possible to realize an arbitrary combinational function as a two level NAND circuit. A two level NAND circuit realization of a function $f$ can be obtained directly from a sum of products realization of $f$ by the sequence of transformations depicted in Figure 3.27.

The circuit of Figure 3.27(b) is obtained from that of Figure 3.27(a) by DeMorgan's Law, $P_1 + P_2 + \ldots + P_n = \overline{\bar{P}_1 \cdot \bar{P}_2 \ldots \bar{P}_n}$. Since a NAND gate is trivially equivalent to an AND gate followed by an inverter, the circuit of Figure 3.27(c) is derived from that of Figure 3.27(b). Note that the inputs to the first level gates have not been changed. Hence simply changing all gates in the sum of products realization of $f$ to

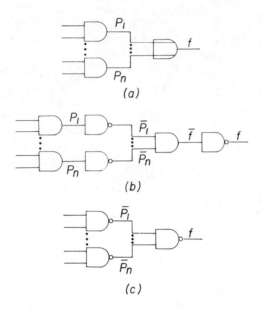

**Figure 3.27   Transformation to NAND realization**

NAND gates results in a two level NAND realization of $f$. The minimal cost two level NAND realization is obtained by applying this transformation to the minimal cost sum of products realization. (The proof of this statement is virtually identical to that of Theorem 3.2 and is left as an exercise.)

A two level NOR gate realization of a combinational function $f$ can be obtained from a product of sums realization of $f$ by the sequence of transformations depicted in Figure 3.28. The circuit of Figure 3.28(b) is obtained from that of Figure 3.28(a) by DeMorgan's Law,

$$P_1 \cdot P_2 \cdot \ldots \cdot P_n = \overline{\bar{P}_1 + \bar{P}_2 + \ldots + \bar{P}_n}.$$

The circuit of Figure 3.28(c) is generated by recognizing that an OR gate followed by an inverter is equivalent to a NOR gate. Note that the inputs to the first level gates have not been changed by this transformation. Thus a product of sums realization of $f$ is converted to a two level NOR realization of $f$ by changing all gates to NOR gates. Applying this tranformation to the minimal cost product of sums realization results in the minimal cost two level NOR realization (Exercise).

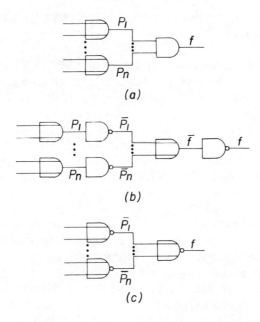

Figure 3.28  Transformation to NOR realization

## Example 3.12

For the function of Example 3.10 minimal NAND and NOR realizations (Figures 3.29) are obtained from the minimal sum of products and product of sums realizations respectively.

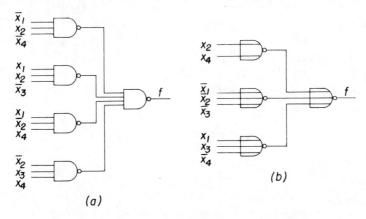

Figure 3.29  NAND and NOR realizations of function of Example 3.10

## 3.3 MULTIPLE OUTPUT COMBINATIONAL CIRCUIT
## MINIMIZATION

In the previous section we developed procedures for the design of
minimal two level realizations of arbitrary single output (i.e. binary valued)
combinational functions. In this section we shall consider the design
of minimal two level realizations of circuits which realize a set $F$
$= \{f_1, f_2, ..., f_r\}$ of binary valued combinational functions. Such circuits
will be called *multiple output combinational circuits*. Assuming that an
AND-OR (sum of products) circuit is to be designed to realize $F$, each
function $f_i$ corresponds to the output of an OR gate, the inputs of which
are AND gates. If the output of each AND gate is only permitted to
be an input of a single OR gate then the problem reduces to realizing
each individual function $f_i$ in a minimal two level realization as in the
previous section. In this case we say that the *sharing* of logic gates
between functions is prohibited. However if logic sharing is permitted
then it becomes necessary to generalize the single function minimization
procedure. In particular the concept of a prime implicant must be
generalized to a *multiple output prime implicant*. Consider the two
functions of Figure 3.30. The multiple output two level AND-OR (sum
of products) circuit realization which minimizes the total number of
gate inputs for this pair of functions is shown in Figure 3.31. The term

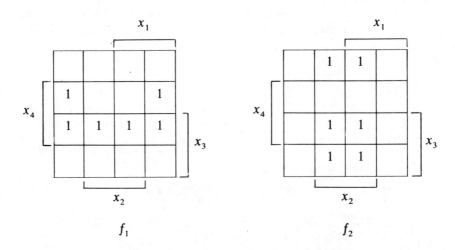

Figure 3.30  A multiple output combinational function

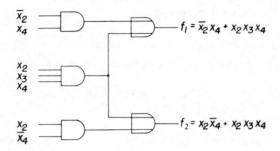

$$f_1 = \overline{x}_2 x_4 + x_2 x_3 x_4$$

$$f_2 = x_2 \overline{x}_4 + x_2 x_3 x_4$$

Figure 3.31 The minimal sum of products realization

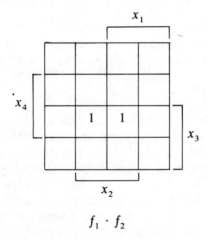

$$f_1 \cdot f_2$$

Figure 3.32 Karnaugh map of $f_1 \cdot f_2$

$x_2 x_3 x_4$ is *not* a prime implicant of either $f_1$ or $f_2$ as we can see from the Karnaugh maps of Figure 3.30. However, it is a prime implicant of the product of these two functions $f_1 \cdot f_2$ as seen from the Karnaugh map of Figure 3.32. We will account for the use of such a product term in a minimal cost circuit by generalizing the concept of prime implicant to *multiple output prime implicant* as follows:

Given a set of combinational functions $F = \{f_1, f_2, ..., f_r\}$, a product term $P$ is a *multiple output prime implicant* of a subset $F'$ of $F$ if $P$ is a prime implicant of $G = \Pi_{f_i \in F'} f_i$ ($G$ is the product of all functions in the subset $F'$). The product of two functions $f_1(x_1, ...,x_n)$ and

$f_2(x_1 \ldots x_n)$ is defined as follows:

$f_1 \cdot f_2(x_i) = 0$ if $f_1(x_i)$ or $f_2(x_i)$ is equal to 0

$f_1 \cdot f_2(x_i) = 1$ if $f_1(x_i)$ and $f_2(x_i)$ are equal to 1

$f_1 \cdot f_2(x_i) = -$ if neither $f_1(x_i)$ nor $f_2(x_i)$ is equal to 0 and $f_1(x_i)$ and/or $f_2(x_i)$ is unspecified.

The following theorem shows that minimal two level AND-OR (sum of product) realizations of sets of combinational functions can be obtained in which the output of each AND gate is a multiple output prime implicant.

**Theorem 3.3:**

Given a set of combinational functions $F = \{f_1, f_2, \ldots, f_r\}$ there exists a multiple output two level sum of products realization which is minimal with respect to total gate inputs and in which the output of every AND gate is a multiple output prime implicant.

**Proof:**

Consider a minimal AND-OR realization of a set of functions $F = \{f_1, f_2, \ldots, f_r\}$ and suppose the output of some AND gate $A$ is a product term $P$ which is not a multiple output prime implicant. Assume the output of $A$ is an input to the OR gates realizing the subset of combinational functions $F' \subseteq F$. Let $G = \Pi_{f_i \in F'} f_i$. Since $P$ is not a multiple output prime implicant of $G$ there must exist a multiple output prime implicant $P'$ of $G$ such that (1) $P'$ covers every 1-point of each $f_i \in F'$ covered by $P$, (2) every point covered by $P'$ is a 1-point or is unspecified for each $f_i \in F'$, and (3) the number of literals in $P'$ does not exceed the number of literals in $P$. Hence we can replace gate $A$ by a gate $A'$ which realizes the multiple-output prime implicant $P'$ and the circuit remains a realization of $F$. This can be repeated until all such AND gates are replaced and the output of every AND gate is a multiple output prime implicant.

It can also be shown that there exists a sum of products realization which is optimal with respect to number of gates in which the output of each AND gate is a multiple output prime implicant.

Thus the problem of generating a minimal two level AND-OR realization for a set of combinational functions reduces to the following general procedure.

**Procedure 3.4**

1) Generate the set of all multiple output prime implicants of $F$ = $\{f_1, f_2, ..., f_r\}$. (Note that prime implicants of an individual function $f_i$ are also multiple output prime implicants where the subset $F'$ contains only $f_i$.)

2) Select a "minimal cost" set of multiple output prime implicants to realize the set of combinational functions $F = \{f_1, f_2, ..., f_r\}$. Such a set must cover all 1-points of each function $f_i \epsilon F$. If a selected term $P$ is a multiple output prime implicant of $G = \Pi_{f_i \epsilon F'} f_i$ then this term will be used as an implicant for each $f_i \epsilon F'$, (i.e. $P$ will be realized by an AND gate, the output of which will be an input to the OR gates realizing each of the functions $f_i \epsilon F'$).

### 3.3.1 GENERATION OF MULTIPLE OUTPUT PRIME IMPLICANTS

The complete set of multiple output prime implicants for a set of combinational functions $F = \{f_1, f_2, ..., f_r\}$ can be determined from the Karnaugh maps of each individual function and the products of all subsets of $F$. Thus $2^r - 1$ Karnaugh maps must be considered in all.

**Example 3.13**

Consider the set of combinational functions $F = \{f_1, f_2, f_3\}$ whose Karnaugh map representations are shown in Figure 3.33(a). The product $f_1 \cdot f_3$ is identical to $f_3$. The product functions $f_1 \cdot f_2, f_2 \cdot f_3$, and $f_1 \cdot f_2 \cdot f_3$ are shown in Figure 3.33(b). From these maps we determine the following complete set of multiple output prime implicants.

The multiple output prime implicants of $f_1$ are $\bar{x}_1 \bar{x}_3, \bar{x}_1 x_2$, $x_2 x_3, x_1 x_3, x_1 \bar{x}_2, \bar{x}_2 \bar{x}_3$. The multiple output prime implicants of $f_2$ are $\bar{x}_1 \bar{x}_2, \bar{x}_2 x_3, x_1 x_3$. The multiple output prime implicants of $f_3$ and $f_1 \cdot f_3$ are $\bar{x}_1 \bar{x}_3, \bar{x}_2 \bar{x}_3, \bar{x}_1 x_2$. The multiple output prime implicants of $f_1 \cdot f_2$ are $\bar{x}_1 \bar{x}_2 \bar{x}_3, x_1 x_3$. There are no multiple output prime implicants of $f_2 \cdot f_3$ and $f_1 \cdot f_2 \cdot f_3$.

Notice that the same term may be a multiple output prime implicant for more than one product of functions. For instance, in the previous example, $\bar{x}_1 \bar{x}_3$ is a multiple output prime implicant of $f_1$ and of $f_1 \cdot f_3$.

Obviously as $r$ becomes large the amount of computation required increases greatly. However the tabular method for generation of prime

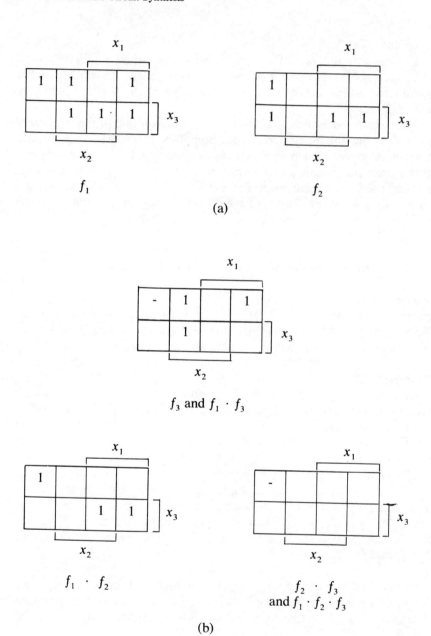

Figure 3.33   Karnaugh maps for Example 3.13

implicants of single combinational functions can be extended to generate multiple output prime implicants for a set of functions $F$ without explicitly generating the product of all subsets of $F$ as follows.

We create a table listing all 1-points and unspecified points of each individual function $f_1, f_2, ..., f_r$ of $F$ and group these points into sets according to the number of 0 and 1 variables as in Procedure 3.2. Associated with each row is a function identifier which indicates for each function of $F$ whether or not the given row is a 0-point of that function. When rows are combined the function identifiers can be combined so as to indicate the set of functions for which the combined term is a multiple output prime implicant. Formally the procedure is as follows.

**Procedure 3.5**

1) Form a table listing all 1-points and unspecified points for the entire set of functions $\{f_1, f_2, ..., f_r\}$. A minterm $m_i$ is represented by the binary representation of $i$ in variables $x_1, x_2, ..., x_n$. Partition this list into sets $S_0, S_1, ..., S_i$ where set $S_i$ consists of all such points having $i$ variables equal to 1 and $(n - i)$ variables equal to 0. Add a set of $r$ columns to the table headed by $f_1, f_2, ..., f_r$. Associate with a row representing a point $x_i$ in the table a function identifier in these $r$ columns with a 0 in column $f_k$ if $f_k(x_i) = 0$, and a 1 in column $f_k$ if $f_k(x_i) \neq 0$.

2) Rows of this table will be merged to form sets $S_i'$ in the same way as for single combinational functions (Procedure 3.2). When rows $x_i$ and $x_j$ are merged the function identifier of the resultant row has a 1 in column $f_k$ if $f_k(x_i) \neq 0$ *and* $f_k(x_j) \neq 0$. If $f_k(x_i)$ or $f_k(x_j) = 0$, the entry in column $f_k$ is 0.

If all $r$ entries of the function identifier are 0 the associated row merger does not represent a multiple output prime implicant and the row can be removed from the table. If the function identifier of the merged row is identical to the function identifier of $x_i$ (and/or $x_j$) then row $x_i$ (and/or $x_j$) are not multiple output prime implicants. This can be denoted by flagging ($\checkmark$) $x_i$ (and/or $x_j$).

3) Continue combining rows in this manner. When no more rows can be combined those unflagged rows with at least one 1-entry in the function identifier columns are multiple output prime implicants* and

*As for Procedure 3.2, some of these multiple output prime implicants may not cover any 1-points. They will be discarded in the covering part of the problem. To avoid generating these prime implicants, a 3-valued function identifier would be required.

the associated set of functions is indicated by the function indicator.

Procedure 3.5 will generate all multiple output prime implicants of a set of functions $F = \{f_1, f_2, ..., f_r\}$ with the following exception. A multiple output prime implicant $P$ of a subset $F'$ of $F$ may also be a multiple output prime implicant of a subset of $F'$, $F''$. Procedure 3.5 will only have one row corresponding to this prime implicant and the associated function indicator will correspond to $F'$. To determine whether $P$ is a multiple output prime implicant of $F''$ examine all unflagged rows of the table which have no 0-entries in any column $f_i \in F''$. If any such row covers all points covered by $P$ then $P$ is not a multiple output prime implicant of $F''$.

**Example 3.14**

For the functions $f_1$, $f_2$, $f_3$ of Figure 3.33, Procedure 3.5 would generate the table shown in Figure 3.34. The sets $S_0'$, $S_1'$, $S_2'$ are generated as in Procedure 3.2 and the function identifiers are generated as specified

| | | $x_1$ | $x_2$ | $x_3$ | $f_1$ | $f_2$ | $f_3$ | |
|---|---|---|---|---|---|---|---|---|
| $S_0$ | { | 0 | 0 | 0 | 1 | 1 | 1 | |
| $S_1$ | { | 0 | 0 | 1 | 0 | 1 | 0 | ✓ |
| | | 0 | 1 | 0 | 1 | 0 | 1 | ✓ |
| | | 1 | 0 | 0 | 1 | 0 | 1 | ✓ |
| $S_2$ | { | 0 | 1 | 1 | 1 | 0 | 1 | ✓ |
| | | 1 | 0 | 1 | 1 | 1 | 0 | ✓ |
| $S_3$ | { | 1 | 1 | 1 | 1 | 1 | 0 | ✓ |
| | | | | | | | | |
| $S_0'$ | { | 0 | 0 | - | 0 | 1 | 0 | |
| | | 0 | - | 0 | 1 | 0 | 1 | |
| | | - | 0 | 0 | 1 | 0 | 1 | |
| $S_1'$ | { | - | 0 | 1 | 0 | 1 | 0 | |
| | | 0 | 1 | - | 1 | 0 | 1 | |
| | | 1 | 0 | - | 1 | 0 | 0 | |
| $S_2'$ | { | - | 1 | 1 | 1 | 0 | 0 | |
| | | 1 | - | 1 | 1 | 1 | 0 | |

Figure 3.34  Tabular generation of multiple output prime implicants

in step (2) of Procedure 3.5. For example the points (100) of $S_1$ and (101) of $S_2$ are combined to form the 1-cube (10−) of $S'_1$. The function identifier of (10−) has a 1 in $f_1$ since $f_1(1,0,0) = 1$ and $f_1(1,0,1) = 1$, a 0 in $f_2$ since $f_2(1,0,0) = 0$ and a 0 in $f_3$ since $f_3(1,0,1) = 0$. Note that point (001) of $S_1$ and (011) of $S_2$ could be combined but the resulting 2-cube (0-1) has a function identifier consisting of all 0's indicating that $\bar{x}_1 x_3$ is not a multiple output prime implicant of any subset of $\{f_1, f_2, f_3\}$.

The term $x_1 x_3$ is indicated to be a prime implicant of $f_1 \cdot f_2$ by the last row of this table. Whether it is a prime implicant of $f_1$ alone (and/or $f_2$ alone) can be determined by examining all prime implicants with a 1 in column $f_1$ (or $f_2$) and determining whether any such prime implicant covers all points covered by $x_1 x_3$. This leads to the conclusion that $x_1 x_3$ is a multiple output prime implicant of $f_1$ and of $f_2$ also. Similarly $\bar{x}_1 \bar{x}_2 \bar{x}_3$ is indicated as a prime implicant of $f_1 \cdot f_2 \cdot f_3$. It is not a prime implicant of $f_1 \cdot f_3$ (since $\bar{x}_1 \bar{x}_3$ covers all points covered by $\bar{x}_1 \bar{x}_2 \bar{x}_3$ and is a prime implicant of $f_1 \cdot f_3$). It is not a prime implicant of $f_2$, (since $\bar{x}_1 \bar{x}_2$ covers all points covered by $\bar{x}_1 \bar{x}_2 \bar{x}_3$ and is a prime implicant of $f_2$). However it is a prime implicant of $f_1 \cdot f_2$ since no other prime implicant with 1's in columns $f_1$ and $f_2$ covers the point (000).

### 3.3.2 SELECTION OF A MINIMAL COVERING SET OF MULTIPLE-OUTPUT PRIME IMPLICANTS

The selection of a minimal covering set of multiple output prime implicants can be formulated as a prime implicant table covering problem. The table has a row for each multiple output prime implicant and a column for each 1-point of each function $f_i$. If column $j$ represents a 1-point $m_r$ of $f_k$, the entry in row $i$ column $j$ is 1 if row $i$ represents a multiple output prime implicant $P$ of $F'$, $f_k$ is contained in $F'$, and $P$ covers $m_r$. As in the case of single function minimization, the cost associated with a prime implicant depends on the optimization goal. To minimize the total number of gates the cost associated with prime implicant $P_i$ containing $q$ literals is 1 if $q > 1$ and 0 if $q = 1$. If the optimization goal is to minimize the total number of gate inputs required, if an implicant $P_i$ with $q$ literals is a multiple output prime implicant of a set of $k$ functions, the cost of $P_i$ is $q + k$ if $q > 1$ (consisting of a $q$-input AND gate and $k$ OR gate inputs). As indicated previously a multiple output prime implicant of $f_i \cdot f_j$ may also be a multiple output prime implicant of $f_i$ alone (and/or $f_j$ alone). In this case $P_i$ would appear in the table two (or three times), once covering the 1-points of $f_i$ and also the 1-points of $f_j$ with cost $q + 2$ and once covering only the 1-points of $f_i$ (and/or covering only the 1-points of $f_j$) with cost $q + 1$.

The table can frequently be simplified by the same type of dominance relations as for the single function case. In addition we can sometimes take advantage of the fact that the same implicant appears several times in the table. Suppose $P_i$ appears as a prime implicant of $f_i$ and of $(f_i \cdot f_j)$ and in order to cover all 1-points of $f_i$ one of these rows must be selected. This implies that in the minimal cost circuit, $P_i$ must be an input to the OR gate realizing $f_i$ but may or may not be shared with $f_j$. Therefore, we can select $P_i$ as a prime implicant of $f_i$ and reduce the table by deleting the row corresponding to $P_i$ as a multiple output prime implicant of $f_i$ and deleting all columns covered by that row.

In reducing the table the cost of the row corresponding to $P_i$ as a multiple output prime implicant of $(f_i \cdot f_j)$ is reduced to 1, since sharing of a product term which has already been selected to be generated requires only one additional gate input. If this row is eventually selected as part of the minimal cover it replaces the use of $P_i$ as a prime implicant of $f_i$ alone. (In effect we select $P_i$ to be generated by an AND gate but defer the decision as to whether $P_i$ will be used only to generate $f_i$ or to generate both $f_i$ and $f_j$.)

**Example 3.15**

For the functions $\{f_1, f_2, f_3\}$ of Figure 3.33, the covering table is as shown in Figure 3.35.

The last two rows of this table (which from Figure 3.33 are seen not to cover any 1-points of $f_2 \cdot f_3$ and $f_1 \cdot f_2 \cdot f_3$ respectively) are both dominated by the multiple output prime implicant $\bar{x}_1 \bar{x}_2 \bar{x}_3$ of $f_1 \cdot f_2$ and hence can be eliminated from the table. From the resulting reduced table we observe that $x_1 x_3$ is the only multiple output prime implicant which covers $m_7$ of $f_2$. However it is a prime implicant of $f_2$ and of $f_1 \cdot f_2$. We select $x_1 x_3$ to cover this 1-point of $f_2$ and reduce the cost of the corresponding multiple output prime implicant of $f_1 \cdot f_2$ to 1. Similarly $\bar{x}_1 x_2$ and $\bar{x}_2 \bar{x}_3$ are the only multiple output prime implicants which cover $m_3$ and $m_4$, respectively of $f_3$. Selecting $\bar{x}_1 x_2$ and $\bar{x}_2 \bar{x}_3$ to realize $f_3$ reduces the cost of the corresponding multiple output prime

| Mult. Out. Prime implicants and assoc. function sets | 1-points of $f_1$ | | | | | | 1-points of $f_2$ | | | | 1-points of $f_3$ | | | | Cost |
|---|---|---|---|---|---|---|---|---|---|---|---|---|---|---|---|
| | $m_0$ | $m_2$ | $m_3$ | $m_4$ | $m_5$ | $m_7$ | $m_0$ | $m_1$ | $m_5$ | $m_7$ | $m_0$ | $m_2$ | $m_3$ | $m_4$ | |
| $f_1$ $\begin{cases} \bar{x}_1\bar{x}_3 \end{cases}$ | 1 | 1 | | | | | | | | | | | | | 3 |
| $\bar{x}_1 x_2$ | | 1 | 1 | | | | | | | | | | | | 3 |
| $x_2 x_3$ | | | 1 | | | 1 | | | | | | | | | 3 |
| $x_1 x_3$ | | | | | 1 | 1 | | | | | | | | | 3 |
| $x_1 \bar{x}_2$ | | | | 1 | 1 | | | | | | | | | | 3 |
| $\bar{x}_2 \bar{x}_3$ | 1 | | | 1 | | | | | | | | | | | 3 |
| $f_2$ $\begin{cases} \bar{x}_1 \bar{x}_2 \end{cases}$ | | | | | | | 1 | 1 | | | | | | | 3 |
| $\bar{x}_2 x_3$ | | | | | | | | 1 | 1 | | | | | | 3 |
| $x_1 x_3$ | | | | | | | | | 1 | 1 | | | | | 3 |
| $f_3$ $\begin{cases} \bar{x}_1 x_3 \end{cases}$ | | | | | | | | | | | | | 1 | | 3 |
| $\bar{x}_2 \bar{x}_3$ | | | | | | | | | | | 1 | | | 1 | 3 |
| $\bar{x}_1 x_2$ | | | | | | | | | | | | 1 | 1 | | 3 |
| $f_1 \cdot f_2$ $\begin{cases} \bar{x}_1 \bar{x}_2 \bar{x}_3 \end{cases}$ | 1 | | | | | | 1 | | | | | | | | 5 |
| $x_1 x_3$ | | | | | 1 | 1 | | | 1 | 1 | | | | | 4 |
| $f_1 \cdot f_3$ $\begin{cases} \bar{x}_1 x_3 \end{cases}$ | | | 1 | | | | | | | | | | 1 | | 4 |
| $\bar{x}_2 \bar{x}_3$ | 1 | | | 1 | | | | | | | 1 | | | 1 | 4 |
| $\bar{x}_1 x_2$ | | 1 | 1 | | | | | | | | | 1 | 1 | | 4 |
| $f_2 \cdot f_3$ $\begin{cases} \bar{x}_1 \bar{x}_2 \bar{x}_3 \end{cases}$ | | | | | | | 1 | | | | 1 | | | | 5 |
| $f_1 \cdot f_2 \cdot f_3$ $\begin{cases} \bar{x}_1 \bar{x}_2 \bar{x}_3 \end{cases}$ | 1 | | | | | | 1 | | | | 1 | | | | 6 |

**Figure 3.35   Prime implicant covering table**

implicants of $f_1 \cdot f_3$ to 1. After table reduction the multiple output prime implicant $\bar{x}_1 \bar{x}_2$ of $f_2$ dominates the multiple output prime implicant $\bar{x}_2 x_3$ of $f_2$. Deleting the latter makes the former essential with respect to $m_1$ of $f_2$. Hence this row is selected and the table is reduced to that shown in Figure 3.36. The optimal cover for this table consists of $\bar{x}_1 x_2, x_1 x_3, \bar{x}_2 \bar{x}_3$ all of which represent product terms which were used in the realization of $f_2$ or $f_3$.

1-points of $f_1$

|  | $m_0$ | $m_2$ | $m_3$ | $m_4$ | $m_5$ | $m_7$ | Cost |
|---|---|---|---|---|---|---|---|
| $\bar{x}_1 \bar{x}_3$ | 1 | 1 |  |  |  |  | 3 |
| $\bar{x}_1 x_2$ |  | 1 | 1 |  |  |  | 1 |
| $x_2 x_3$ |  |  | 1 |  |  | 1 | 3 |
| $x_1 x_3$ |  |  |  |  | 1 | 1 | 1 |
| $x_1 \bar{x}_2$ |  |  |  | 1 | 1 |  | 3 |
| $\bar{x}_2 \bar{x}_3$ | 1 |  |  | 1 |  |  | 1 |

Figure 3.36    Reduced covering table

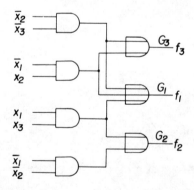

Figure 3.37    Minimal sum of products realization

This results in the optimal sum of products realization for the set of functions $F = \{f_1, f_2, f_3\}$ shown in Figure 3.37, requiring 15 gate inputs. Note that all inputs of the OR gate $G_1$ are also inputs of OR gates $G_2$ or $G_3$.

A minimal cost product of sums (OR-AND) realization of $F = \{f_1, f_2, ..., f_r\}$ could be obtained from the minimal cost sum of products realization of $\bar{F} = \{\bar{f}_1, \bar{f}_2, ..., \bar{f}_r\}$ by DeMorgans Law as in the single function minimization. (Exercise) In general the cost of the minimal product of sums realization and the minimal sum of products realization will be different.

As for the single function case, minimal covering sets for a set of functions $F = \{f_1, f_2, ..., f_r\}$ can sometimes be obtained directly from the Karnaugh maps of the individual functions and the maps of the products of all subsets of these functions. This is done by selecting *essential* multiple output prime implicants, (a multiple output prime implicant $P$ is essential with respect to a 1-point $m_i$ of a function $f_j \epsilon F$ if $P$ is the only multiple output prime implicant which covers $m_i$ of $f_j$), or *good* multiple output prime implicants. A multiple output prime implicant $P$ is *good* with respect to a 1-point $m_i$ of a function $f_j \epsilon F$ if for any other multiple output prime implicant $P'$ which covers $m_i$ of $f_j$:

(1) $P$ covers all previously uncovered 1-points of all functions in $F$ which are covered by $P'$ and

(2) the cost of $P \leq$ the cost of $P'$ where the cost is as previously defined.

If a multiple output prime implicant $P$ is essential or good with respect to a 1-point $m_i$ of $f_j$ then $P$ is selected as part of the cover for $f_j$ and all 1-points of $f_j$ (and of all product functions containing $f_j$) which are covered by $P$ are changed to don't care entries. Such a prime implicant may or may not be used to realize another function $f_k \epsilon F$. In evaluation of good prime implicants it must be remembered that if any previously selected prime implicant $P'$ covers the relevant 1-point $m_i$ of the function under consideration $f_j$, if $P'$ is an implicant of $f_j$ then its cost is 1 since it has been previously selected (assuming gate input minimization). Thus the effect of previously selected prime implicants cannot be ignored.

If a multiple output prime implicant $P$ is essential, then $P$ must be essential for some individual function $f_i$ in $F$. Therefore to discover essential multiple output prime implicants we might first wish to consider essential prime implicants of the individual functions. Also if for some point $\mathbf{x}_k$, $f_i(\mathbf{x}_k) = 1$ and $f_j(\mathbf{x}_k) = 0$ for all $j \neq i$, then this 1-point can only be covered by prime implicants of $f_i$. We might wish to attempt to find prime implicants which are good or essential with respect to such a point. Guidelines such as these facilitate the solution of minimal cost multiple output combinational circuit realizations directly from the Karnaugh maps.

**Example 3.16**

We will find a minimal gate-input two level realization for the set of functions $F = \{f_1, f_2\}$ of Figure 3.38. We first determine the minimal sum of products realization. From the Karnaugh maps of Figure 3.38, $x_2 x_3 x_4$ is essential with respect to $m_{15}$ of $f_2$, and $\bar{x}_2 x_3 \bar{x}_4$ is essential with respect to $m_{10}$ of $f_1$. Also $\bar{x}_1 x_2 x_4$ is good with respect to the 1-point (0101) of $f_1$, since the only other prime implicant which covers $m_5$ of $f_1$ is $\bar{x}_1 \bar{x}_3 x_4$ and that does not cover any other 1-points of $f_1$ and is not an implicant of $f_2$. Note that $\bar{x}_1 x_2 x_4$ is not good with respect

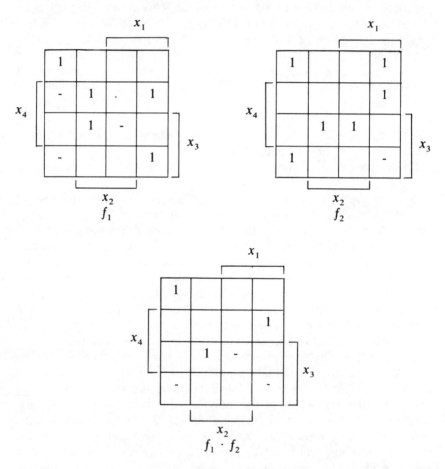

Figure 3.38    Karnaugh maps for Example 3.16

to (0111) of $f_1$ since that point is also covered by the previously selected term $x_2 x_3 x_4$ which is also an implicant of $f_1$ and has cost 1 (since it was previously selected). Similarly $\bar{x}_1 \bar{x}_2 \bar{x}_4$ is good with respect to $m_0$ of $f_1$ since the only other multiple output prime implicant which covers that 1-point is $\bar{x}_1 \bar{x}_2 \bar{x}_3$ which does not cover any other 1-points of $f_1$ and is not an implicant of $f_2$. Selecting these 4 terms and changing all 1-points covered by them to don't cares, results in the Karnaugh maps of Figure 3.39. From these maps it appears that the term $\bar{x}_2 \bar{x}_4$ is a good multiple output prime implicant with respect to $m_2$ of $f_2$.

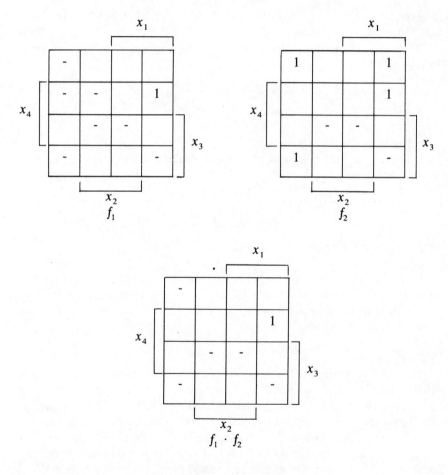

Figure 3.39  Simplified maps

However this point is also covered by $\bar{x}_1 \bar{x}_2 \bar{x}_4$ which is an implicant of $f_2$ and since $\bar{x}_1 \bar{x}_2 \bar{x}_4$ has been previously selected, it has cost 1. Therefore $\bar{x}_2 \bar{x}_4$ is not good. We will therefore branch around these two possible choices to cover $m_2$ of $f_2$. Selecting $\bar{x}_2 \bar{x}_4$ to cover $m_2$ of $f_2$ leads to the minimal gate-input solution

$$f_1 = \bar{x}_1 x_2 x_4 + \bar{x}_1 \bar{x}_2 \bar{x}_4 + \bar{x}_2 x_3 \bar{x}_4 + \underline{x_1 \bar{x}_2 \bar{x}_3 x_4}.$$
$$f_2 = x_2 x_3 x_4 + \bar{x}_2 \bar{x}_4 + \underline{x_1 \bar{x}_2 \bar{x}_3 x_4}$$

where the shared term $x_1 \bar{x}_2 \bar{x}_3 x_4$ is underlined. This solution requires 25 gate inputs.

If $\bar{x}_1 \bar{x}_2 \bar{x}_4$ is used to cover $m_2$ of $f_2$ the optimal solution obtained is

$$f_1 = \bar{x}_1 x_2 x_4 + \underline{\bar{x}_1 \bar{x}_2 \bar{x}_4} + \bar{x}_2 x_3 \bar{x}_4 + \bar{x}_2 \bar{x}_3 x_4$$
$$f_2 = x_2 x_3 x_4 + \underline{\bar{x}_1 \bar{x}_2 \bar{x}_4} + x_1 \bar{x}_2 \bar{x}_3$$

which also requires 25 gate inputs.

However the minimal product of sums realization of $F$ is

$$f_1 = (\bar{x}_2 + x_4)(x_2 + \bar{x}_3 + \bar{x}_4)(\bar{x}_1 + x_3 + x_4)(\bar{x}_1 + \bar{x}_2 + x_3)$$
$$f_2 = (\bar{x}_2 + x_4)(x_2 + \bar{x}_3 + \bar{x}_4)(\bar{x}_1 + \bar{x}_2 + x_3)(x_1 + x_3 + \bar{x}_4)$$

which requires 22 gate inputs. (Exercise)

In this chapter we have presented procedures for generating minimal two level combinational circuits for two different criteria of optimality. In the next chapter we shall consider synthesis of combinational circuits with more than two levels of logic.

## SOURCES
The tabular method for generation of prime implicants was developed by Quine [10,11] and modified by McCluskey [7]. The map method for prime implicant generation was originated by Veitch [12] and modified by Karnaugh [4]. Procedures for reducing and simplifying covering tables have been presented by McCluskey [6,7], Petrick [9], Gimpel [3], Luccio [5] and Friedman [2]. The material on minimal multiple output realizations is due to Bartee [1], and McCluskey and Schorr [8].

# REFERENCES

[1] Bartee, T. C., "Computer Design of Multiple Output Logical Networks," *IRE Trans. Electronic Computers*, vol. EC-10, pp. 21-30, March 1961.

[2] Friedman, A. D., "Comment on 'A Method for the Selection of Prime Implicants'," *IEEE Transactions on Electronic Computers*, vol. EC-16, pp. 221-222, April 1967.

[3] Gimpel, J. F., "A Reduction Technique for Prime Implicant Tables," *IEEE Transactions on Electronic Computers*, vol. EC-14, pp. 535-541, August 1965.

[4] Karnaugh, M., "The Map Method for Synthesis of Combinational Logic Circuits," *Trans. AIEE*, pt. I, vol. 72, no. 9, pp. 593-598, 1953.

[5] Luccio, F., "A Method for the Selection of Prime Implicants," *IEEE Transactions on Electronic Computers*, vol. EC-15, pp. 205-212, April 1966.

[6] McCluskey, E. J., Jr., *Introduction to the Theory of Switching Circuits*, McGraw-Hill, New York, N.Y., 1965.

[7] McCluskey, E. J., Jr., "Minimization of Boolean Functions," *Bell System Technical Journal*, vol. 35, no. 6, pp. 1417-1444, November 1956.

[8] McCluskey, E. J., Jr. and H. Schorr, "Essential Multiple-Output Prime Implicants," *Mathematical Theory of Automata*, Proceedings Polytechnic Institute of Brooklyn Symposium, vol. 12, pp. 437-457, April 1962.

[9] Petrick, S. R., "On the Minimization of Boolean Functions," *Proc. Symposium on Switching Theory*, ICIP, Paris, France, June 1959.

[10] Quine, W. V., "The Problem of Simplifying Truth Functions," *Am. Math. Monthly*, vol. 59, no. 8, pp. 521-531, October 1952.

[11]  Quine, W. V., "A Way to Simplify Truth Functions," *Am. Math. Monthly*, vol. 62, no. 9, pp. 627–631, November 1955.

[12]  Veitch, E. W., "A Chart Method for Simplifying Truth Functions," *Proc. ACM*, Pittsburgh, Pa., pp. 127–133, May 2-3, 1952.

## PROBLEMS

**3.1)** Design two level minimal gate-input NAND and NOR realizations for the following functions

(a)   $f(x_1,x_2,x_3,x_4) = \sum m_3,m_4,m_5,m_7,m_9,m_{13},m_{14},m_{15}$

(b)   $f(x_1,x_2,x_3,x_4) = \sum m_2,m_5,m_6,m_8,m_{12},m_{15}$

$$+ \sum_{\text{don't cares}} m_1,m_3,m_{11},m_{14}$$

(c)   $f(x_1,x_2,x_3,x_4) = \sum m_3,m_4,m_6,m_8,m_{12},m_{13},m_{14}$

(d)   $f(x_1,x_2,x_3,x_4) = \sum m_0,m_2,m_7,m_8,m_9,m_{10},m_{13}$

$$+ \sum_{\text{don't cares}} m_3,m_6,m_{11},m_{15}$$

**3.2)**  We wish to design a combinational circuit to operate in the following manner.

a) The circuit has four inputs $(x_1,x_2,x_3,x_4)$ and 1 output $(z)$

b) If three or more of the inputs are 1 then $z = 1$ unless $x_1 = 0$

c) If $x_1 = 0$ and exactly two of the inputs are 1 then $z = 0$

d) If exactly one input is 1 then $z = 1$ unless $x_2 = 1$

e) If $x_1 = 1$ and one other input is 1 then $z = 0$

f) If all inputs are 0 then $z = 1$

g) All other input conditions are don't cares

Derive a truth table for the function and express $f$ as a product of *maxterms* with all unspecified entries interpreted as 1's.

**3.3)**  a) For the following combinational function find a minimal two level *NOR* realization.

b) Is there any prime implicate (prime implicant of $\bar{f}$) not included in your solution?

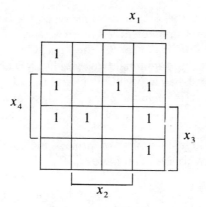

Figure 3.40  Problem 3.3

**3.4)** a) For the combinational function of Figure 3.41 the entries indicated by *a* and *b* are *partial don't cares* in that *a* may be equal to 0 or 1 but *b* must take the opposite value. Thus either point is a don't care but when one of them is specified the other point is no longer a don't care. Find a minimal gate-input two level AND-OR (sum of products) realization.

b) What is the optimal solution if both *a* and *b* are don't cares?

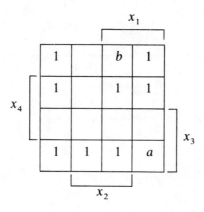

Figure 3.41  Problem 3.4

**3.5)** Assume that information is transferred in a digital system using a 4-bit code in which 3 bits are information bits (which can have

any value) and the 4$^{th}$ bit is a parity check bit which is chosen
so that the total number of 1 bits in a word is odd. For instance
if the information bits are 101 the check bit is 1 resulting in the
word 1011. Design a minimal two level NAND circuit whose output
is 1 if the word has a single bit in error. Will your circuit detect
errors in two bits? Characterize the class of errors your circuit
will detect.

*3.6) What is the largest number of prime implicants a combinational
function of $n$ variables can have.

3.7) a) A function $f(x_1,x_2, ...,x_n)$ is <u>self dual if the</u> dual of $f, f_d$, is
such that $f = f_d$ where $f_d = \overline{f(\bar{x}_1,\bar{x}_2, ...,\bar{x}_n)}$. Prove that the
number of self dual functions of $n$ variables is $2^{2^{n-1}}$.
   b) For the function in the Karnaugh map of Figure 3.42 if possible
   specify values for the don't care entries so that the function
   is *self dual*.

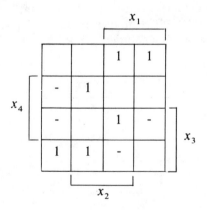

Figure 3.42   Problem 3.7

3.8) a) For the circuit of Figure 3.43 if $f_1 = A + B$, $f_3$ is an AND
gate, and $Z = A$, is $f_2$ uniquely determined?
   b) If $Z = A$ can you specify a function $f_3$ which depends on
   both inputs and enables $f_2$ to be uniquely determined?

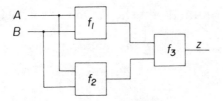

**Figure 3.43   Problem 3.8**

**3.9)** Prove or give a counterexample for each of the following statements.

a) For every prime implicant $P$ of $f(x_1, \ldots, x_n)$ there exists a minimal two level realization of $f$ containing $P$.

b) For a completely specified combinational function there is a unique minimal sum of products expression.

c) For a completely specified combinational function the costs of the minimal sum of products and product of sums expressions are equal.

d) If a combinational function $f(x_1, x_2, \ldots, x_n)$ has a unique minimal sum of products expression then it also has a unique minimal product of sums expression and the two expressions have equal cost.

**3.10)** Use a Karnaugh map to find a minimal two level realization of the function

$$f(x_1, x_2, x_3, x_4) = \prod M_0, M_4, M_6, M_7, M_{10}, M_{12}, M_{13}, M_{14}$$

**3.11)** Use the Quine-McCluskey method to find a minimal two level sum of products realization of the function

$$f(x_1, x_2, x_3, x_4) = \sum_{\text{1-points}} m_4, m_5, m_7, m_{12}, m_{14}, m_{15}$$

$$+ \sum_{\text{don't cares}} m_3, m_8, m_{10}$$

**3.12)** Find minimal NAND or NOR realizations for the following function.

$$f(x_1,x_2,x_3,x_4,x_5) = \sum m_2,m_3,m_7,m_8,m_9,m_{12}, m_{13},m_{14},$$
$$m_{18},m_{19},m_{21},m_{28},m_{29},m_{30}$$

**3.13)** A circuit receives two 3-bit binary numbers $A = A_2 A_1 A_0$, $B = B_2 B_1 B_0$. Design a minimal sum of products circuit to produce an output $z = 1$ if and only if $A$ is greater than $B$.

**3.14)** Five men vote in a contest. Their votes are indicated on inputs $x_i, i = 1, \ldots, 5$ to a logic circuit $C$. If the vote is 5-0 or 4-1 to pass ($x_i = 1$) the circuit outputs should be $z_1 z_2 = 11$. If the vote is 5-0 or 4-1 to fail the circuit outputs should be $z_1 z_2 = 00$. If the vote is 3-2 or 2-3 the output should be $z_1 z_2 = 01$. Design $C$ to be a minimal sum of products circuit.

**3.15)** For the function shown in the Karnaugh map of Figure 3.44 you are permitted to change any single entry to a don't care. Do so in such a manner as to produce the simplest two level realization.

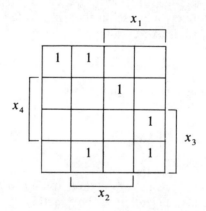

Figure 3.44   Problem 3.15

**3.16)** Define the *majority function* $M(x,y,z) = xy + yz + xz$. Prove or disprove the following.
a) $M(x_1,y_1,z_1) + M(x_2,y_2,z_2) = M(x_1 + x_2, y_1 + y_2, z_1 + z_2)$
b) $M(x_1,y_1,z_1) \cdot M(x_2,y_2,z_2) = M(x_1 \cdot x_2, y_1 \cdot y_2, z_1 \cdot z_2)$
c) $M(M(a,b,c),d,e) = M(M(a,d,e),b,M(c,d,e))$

**3.17)** Define the function $S_i(x_1, x_2, ..., x_n)$ to have the value 1 if and only if exactly $i$ of the $n$ input variables have the value 1.
a) How many prime implicants does $S_i$ have?
b) How many essential prime implicants does $S_i$ have?

**3.18)** Define the function $T_i(x_1, x_2, ..., x_n)$ to have the value 1 if and only if $i$ or more of the input variables have the value 1.
a) How many prime implicants does $T_i$ have?
b) How many essential prime implicants does $T_i$ have?

**3.19)** a) Specify a truth table for the following
(i) There are two functions $f(x_1, x_2, x_3, x_4)$ and $g(x_1, x_2, x_3, x_4)$
(ii) $f = 1$ when 2 or more of the input signals are 1, otherwise $f = 0$.
(iii) $g = 1$ when the number of inputs with a 1 applied is even; in all other cases $\bar{f} = g$.
b) Find minimal sum-of-products and product-of-sums expressions for the functions $f$ and $g$.

**3.20)** For each of the following sets of functions find a minimal 2-level realization.

a) $f_1(x_1, x_2, x_3) = \sum m_0, m_2, m_3, m_6$

   $f_2(x_1, x_2, x_3) = \sum m_2, m_3, m_4, m_6$

b) $f_1(x_1, x_2, x_3) = \sum m_0, m_3 + \sum_{\text{don't cares}} m_2, m_7$

   $f_2(x_1, x_2, x_3) = \sum m_2, m_3, m_4 + \sum_{\text{don't cares}} m_7$

c) $f_1(x_1, x_2, x_3, x_4) = \sum m_5, m_7, m_{12}, m_{13} + \sum_{\text{don't cares}} m_2$

   $f_2(x_1, x_2, x_3, x_4) = \sum m_0, m_1, m_2, m_5 + \sum_{\text{don't cares}} m_7$

   $f_3(x_1, x_2, x_3, x_4) = \sum m_1, m_2, m_5, m_{12} + \sum_{\text{don't cares}} m_{13}$

**3.21)** Find a minimal sum of products using a Karnaugh map and a minimal product of sums using a table technique for the following functions, considered individually and considered as a multi-output circuit.

$$f(x_1,x_2,x_3,x_4) = \sum m_1,m_4,m_5,m_6,m_{11},m_{12},m_{13},m_{14},m_{15}$$
$$g(x_1,x_2,x_3,x_4) = \sum m_0,m_2,m_4,m_9,m_{12},m_{15}$$
$$+ \sum_{\text{don't cares}} m_1,m_5,m_7,m_{10}$$

**3.22)** Consider a digital system in which all registers have $n$ bits, and addition and subtraction are done in 2's complement form. If the addition of $X = x_{n-1}\, x_{n-2} \cdots x_0$ and $Y = y_{n-1}\, y_{n-2} \cdots y_0$ results in an output $Z = z_{n-1}\, z_{n-2} \cdots z_0$ then if $X + Y \geq 2^{n-1}$ or $X + Y \leq -2^{n-1}$ an overflow occurs. Design a simple circuit which generates an output $g = 1$ if overflow has occurred.

**3.23)** Consider the ring sum function $A \oplus B = A\bar{B} + \bar{A}B$, where $A$ and $B$ are Boolean expressions.
   a) Show that this function is commutative and associative.
   b) Show that the function $x_1 \oplus x_2 \oplus ... \oplus x_n$ has the value 1 if and only if an odd number of input variables are 1.
   c) Given a function $f(x_1, ...,x_n) = \Sigma_{i \epsilon I}\, m_i$ show that $f = \Sigma_{\oplus i \epsilon I}\, m_i$ (Ring sum of minterms is equal to logical OR of minterms).
   d) Show that $1 \oplus A = \bar{A}$
   e) Show that $A(B \oplus C) = AB \oplus AC$
   f) Show that $A \oplus A = 0$
   g) Prove that any completely specified combinational function can be expressed uniquely in the following canonical form.

$$f(x_1,x_2, ...,x_n) = a_0 \oplus a_1 x_1 \oplus a_2 x_2 \oplus ... \oplus a_n x_n$$
$$\oplus\, a_{n+1} x_1 x_2 \oplus a_{n+2} x_1 x_3 \oplus ...$$
$$\oplus\, a_{2^n-1} x_1 x_2...x_n$$

where $a_i = 0$ or 1. (There is one term for each subset of

the variables $(x_1, \ldots, x_n)$. This representation is referred to as the Reed Muller* representation.)

† **3.24)** Prove or present a counterexample to the following: For completely specified functions a minimum gate-input sum of products realization is also a minimum gate sum of products realization.

*Muller, D. E., "Application of Boolean Algebra to Switching Design and to Error Detection," *IRE Transactions on Electronic Computers*, vol. EC-3, pp. 6–12, September 1954.
Reed, I. S., "A Class of Multiple-Error-Correcting Codes and the Decoding Scheme," *IRE Transactions on Information Theory*, vol. IT-4, pp. 38–49, 1954.

# CHAPTER 4

# Advanced Concepts of Combinational Circuit Design

## 4.1 MULTIPLE LEVEL COMBINATIONAL CIRCUITS

We have so far only considered two level realizations of combinational functions. As mentioned previously, if a delay is associated with each gate in a circuit, the circuit speed is proportional to the number of levels. Hence two level circuits have the advantage of being relatively fast. However there are several practical reasons for sometimes designing circuits with more than two levels. (We shall refer to such circuits as *multiple level circuits*). We shall now consider some of these reasons.

For a circuit $C$ *the fanout index* of an input $x_i$ or a gate output $G$ is equal to the number of gates with inputs of $x_i$ or $G$. The *fanout index* of $C$ is the maximum fanout of any input or gate output. The *fan-in index* of a gate $G$ is the number of inputs that $G$ has. The *fan-in index* of $C$ is the maximum fan-in of any gate in $C$. Circuits frequently have constraints on fan-in and fanout indices due to electrical considerations. Two level circuits frequently have relatively high fan-in and fanout indices. However both of these indices can be reduced by increasing the number of levels in the circuit.

An AND (OR) gate with $n$ inputs can be realized as a multi-level circuit of 2-input AND (OR) gates. Such a circuit has $\lceil \log_2 n \rceil$ levels with $n - 1$ AND (OR) gates and fanout index of 1. For $n = 8$ the circuit is as shown in Figure 4.1. It follows from this that any combinational function can be realized by a circuit with fan-in index equal 2.

Similarly any input or gate with fanout index of $n$ can be realized

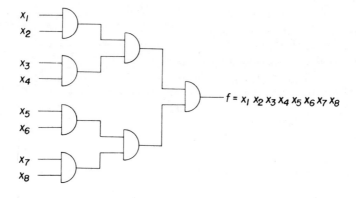

Figure 4.1    Multi-level reduced fan-in gate equivalent

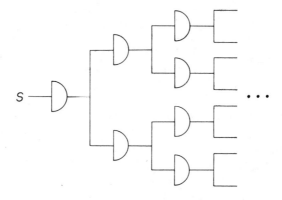

Figure 4.2    Multi-level reduced fan-out gate equivalent

as a circuit with $\lceil \log_2 n \rceil$ levels consisting of 1-input AND gates with fanout index equal to 2. Such a circuit is shown in Figure 4.2.

Thus all combinational functions can be realized as circuits with fanout index at most 2. Circuits with fanout index 1 are sometimes referred to as *fanout free circuits*. Hayes[7] has derived a procedure to determine whether a given combinational function can be realized as a fanout free circuit. We shall consider this problem in Section 4.4 of this chapter.

Multiple level circuits will also frequently be less costly in terms of gates or gate inputs than equivalent 2-level circuits. A striking example of this trade off between logical complexity and speed (i.e., number of levels) is the parity check function, $f(x_1,x_2...x_n) = x_1 \oplus x_2 \oplus ... \oplus x_n$ (where the $\oplus$ operator is as defined in Problem 3.23). This function has the value 1 if an odd number of input variables are 1 and has the value 0 if an even number of input variables have the value 1. The Karnaugh map of this function looks like a checkerboard in that no cell has the same value as any of its neighbors. For four variables the function is as shown in the Karnaugh map of Figure 4.3. A 2-level

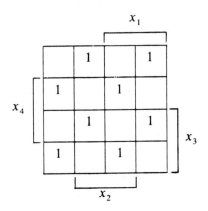

Figure 4.3  Karnaugh map of parity check function

sum of products realization of an $n$-variable parity function has $2^{n-1}$ $n$-input AND gates and a single $2^{n-1}$-input OR gate. This function can also be realized by connecting $(n - 1)$ 2-input parity check circuits in the form of a tree with $\lceil \log_2 n \rceil$ levels of these 2-input modules as shown in Figure 4.4. Such a circuit has $3 \times \lceil \log_2 n \rceil$ levels but only has $3(n - 1)$ 2-input gates and $2(n - 1)$ inverters, and has fanout and fan-in indices equal to two.

### 4.1.1 FACTORIZATION

One method of obtaining multiple level realizations is by *factorization*. This consists of factoring out subexpressions which are common to several different terms in a Boolean expression of $f$, the function to be realized. For example, in the expression $f = x_1 x_2 x_3 \bar{x}_4 + x_1 x_2 \bar{x}_3 x_4$ both terms have a common factor $x_1 x_2$. Thus this Boolean expression

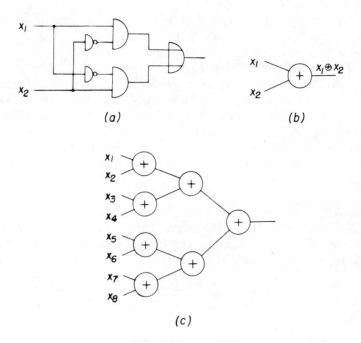

Figure 4.4   (a) A 2-input parity check cell, (b) A schematic representation of a 2-input parity check cell, (c) A multi-level realization of an 8 bit parity check function

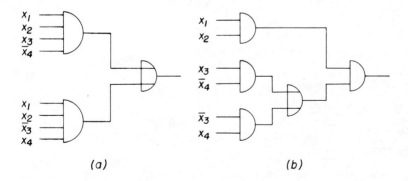

Figure 4.5   (a) A 2-level combinational circuit (b) An equivalent multi-level combinational circuit

can be rewritten as $f = x_1 x_2 (x_3 \bar{x}_4 + \bar{x}_3 x_4)$. The realizations of these two equivalent expressions are shown in Figure 4.5. The factored realization has fan-in index two as compared to four in the original circuit, but the number of levels has increased from two to three.

Of course, factorization can be applied iteratively to an expression. However, large expressions may have many possible factored forms and these forms may have circuit realizations with substantially different properties. In addition, the results obtainable by factoring can often be improved by employing other methods in addition to factoring. These limitations of the factoring procedure are exhibited in the following example.

**Example 4.1**

We will attempt to factor the expression $f = x_1 x_2 + x_2 x_3 + x_1 \bar{x}_3 x_4$ so as to obtain a good realization with fan-in index equal to two. There are two apparent factorizations

$$f = x_1 (x_2 + \bar{x}_3 x_4) + x_2 x_3 \tag{1}$$

and

$$f = x_2 (x_1 + x_3) + x_1 \bar{x}_3 x_4. \tag{2}$$

Of these only (1) satisfies the fan-in index constraint. However, there exists another factorization $f = (x_1 + x_3)(x_2 + \bar{x}_3 x_4)$ (Figure 4.6(b)) which is superior to (1) (Figure 4.6(a)) with respect to gates, gate inputs, and

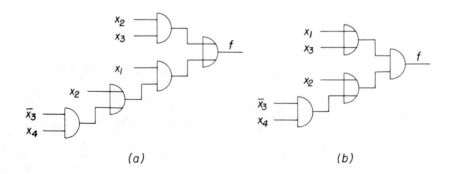

*(a)*                                         *(b)*

Figure 4.6   Two equivalent multi-level circuits

number of levels. This factorization is obtained by adding the superfluous term $x_3\bar{x}_3x_4$ to (1) as follows:

$$f = x_1(x_2 + \bar{x}_3x_4) + x_2x_3$$
$$= x_1(x_2 + \bar{x}_3x_4) + x_2x_3 + x_3\bar{x}_3x_4$$
$$= x_1(x_2 + \bar{x}_3x_4) + x_3(x_2 + \bar{x}_3x_4)$$
$$= (x_1 + x_3)(x_2 + \bar{x}_3x_4)$$

## Decoders

One common type of circuit for which economical multiple level circuits can be systematically designed using factorization techniques is the (*complete*) *decoder*.

Such a circuit has $n$ inputs $x_1,x_2...x_n$ and $2^n$ outputs $z_0,z_1...z_{2^n-1}$ where $z_i = m_i$ (minterm $m_i$ in the $n$-dimensional space defined by $x_1,x_2...x_n$).

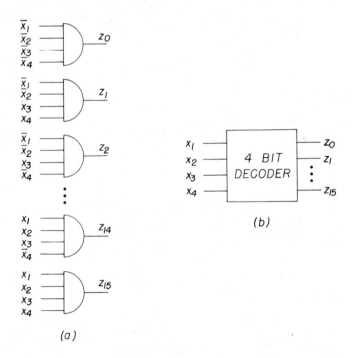

Figure 4.7   One level 4-bit decoder and its schematic representation

Any input to a decoder will result in exactly one 1-output. Such a circuit is sometimes referred to as a *1-hot circuit* for this reason.

An $n$-bit decoder can be realized as a one level circuit consisting of $2^n$ $n$-input AND gates. However, this cost can be considerably reduced by realizing the decoder as a multiple level circuit. Consider the design of a 4-bit decoder. The single level 4-bit decoder circuit is shown in

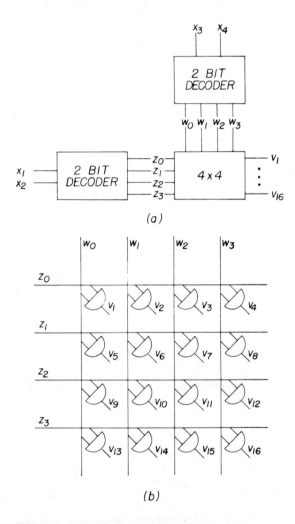

Figure 4.8  (a) Multi-level 4-bit decoder, (b) AND matrix

Figure 4.7(a) which is represented schematically as shown in Figure 4.7(b). This circuit is easily generalized to an $n$-bit decoder having $2^n$ $n$-input AND gates. However, the total number of gate inputs is considerably reduced by realizing the 4-bit decoder from two 2-bit decoders as shown in Figure 4.8(a). The circuit labelled $(4 \times 4)$ is as

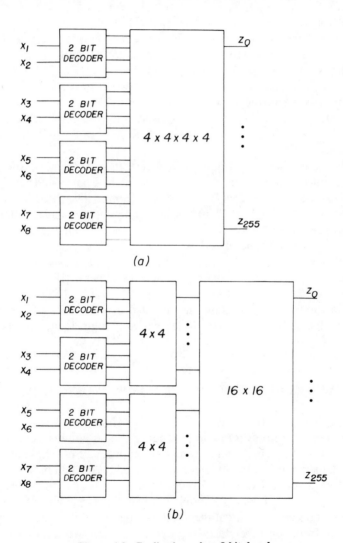

Figure 4.9   Realizations of an 8-bit decoder

shown in Figure 4.8(b). This type of circuit is called an *AND matrix* (or *array*). We shall just refer to it as a matrix. It can be generalized to an $n_1 \times n_2 \times ... \times n_k$ matrix which has $n_1 \cdot n_2 \cdot ... \cdot n_k$ $k$-input gates, each such gate defining the output of a 1-level circuit. The realization of Figure 4.8(a) is effectively obtained by factoring out common subexpressions $(x_1 x_2, x_1 \bar{x}_2, \bar{x}_1 x_2, \bar{x}_1 \bar{x}_2, x_3 x_4, x_3 \bar{x}_4, \bar{x}_3 x_4, \bar{x}_3 \bar{x}_4)$ which are functions of the variable set $\{x_1, x_2\}$ or $\{x_3, x_4\}$.

The one level 4-bit decoder (Figure 4.7(a)) has a total of 64 gate inputs whereas the two level 4-bit decoder (Figure 4.8(a)) has 48 gate inputs. A one level $2p$-bit decoder requires $2p \times 2^{2p}$ gate inputs. This circuit can also be decomposed into two $p$-bit decoders and a $(2^p \times 2^p)$ matrix requiring $(2 \times 2^p \times p) + (2^p \times 2^p) \times 2 = p2^{p+1} + 2^{2p+1}$ gate inputs. It can also be decomposed into $p$ 2-bit decoders and a $\underbrace{4 \times 4 \times ... \times 4}_{p \text{ times}}$ matrix requiring $4p + 4^p \times p = p(4^p + 1)$ gate inputs.

This circuit is shown in Figure 4.9(a), for $p = 4$. This circuit can be factored into a three level circuit as shown in Figure 4.9(b). The two level circuit has $4(8) + 256(4) = 1056$ gate inputs. The three level circuit has $4(8) + 2(4 \times 4)2 + (256)2 = 608$ gate inputs while a one level 8-bit decoder would have $2^8 \times 8 = 2048$ gate inputs. The circuit of Figure 4.9(b) can be generalized to realize a $2^k$-bit decoder, using repeated factorization as a $k$ level circuit with $2^{k-1}$ 2-bit decoders, $2^{k-2}$ $4 \times 4$ matrices, $2^{k-3}$ $16 \times 16$ matrices, and in general $2^{k-i}(4^{i-1} \times 4^{i-1})$ matrices for $2 \le i \le k$, thus requiring $2^{k+2} + \Sigma_{i=2}^{k} 2^{k-i}(16^{2i-2})2$ gate inputs in total. Thus decoding circuits also offer a trade-off possibility between speed (i.e., number of levels) and logical complexity (total number of gate inputs).

## 4.2 MODULAR REALIZATIONS OF COMBINATIONAL FUNCTIONS

Although it may be necessary to realize functions as multiple level circuits there are no systematic procedures for obtaining good multiple level realizations, other than decomposition procedures which are essentially exhaustive in nature and heuristic procedures which may sometimes fail to find good realizations. However, many circuits which frequently occur in practical systems can be realized in a regular manner from interconnections of identical (or similar) smaller modules. Such realizations can frequently be derived directly from a word description of the function they perform without generating a truth table, and in such a way that they define a circuit (or family of circuits) which is relatively independent (in structure) of the number of input variables upon which the function is computed. Such realizations are said to be *modular*. The realization of the parity check function of Figure 4.4 is such a circuit. We shall now consider two other examples of functions which have such a multilevel modular decomposition.

### 4.2.1 A COMPARISON CIRCUIT

Consider the design of a comparison circuit (*comparator*) whose inputs are two numbers $X$ and $Y$, and whose output $Z$, is equal to 1 if $X > Y$ and is equal to 0 if $X \leq Y$. If $X$ and $Y$ are single bit numbers, then the output, $Z$, can be defined by the equation $Z_1(x_o, y_o) = x_o \bar{y}_o$. If $X$ and $Y$ are 2-bit numbers $X = x_1 x_o$, $Y = y_1 y_o$) then the output $Z_2$ can be expressed as a sum of products $Z_2 = x_1 \bar{y}_1 + x_1 y_1 x_o \bar{y}_o + \bar{x}_1 \bar{y}_1 x_o \bar{y}_o$. Factoring out the common term $x_o \bar{y}_o$, $Z_2 = x_1 \bar{y}_1 + x_o \bar{y}_o (x_1 y_1 + \bar{x}_1 \bar{y}_1)$ which can be expressed in terms of the single bit comparison function $Z_1$, as

$$Z_2(x_1, x_0, y_1, y_0) = Z_1(x_1, y_1) + Z_1(x_0, y_0)(x_1 y_1 + \bar{x}_1 \bar{y}_1)$$

$$= Z_1(x_1, y_1) + Z_1(x_0, y_0) E_1(x_1, y_1)$$

where $\qquad E_1(x_1, y_1) = x_1 y_1 + \bar{x}_1 \bar{y}_1.$

Similarly a 3-bit comparison circuit can be simplified as follows

$$Z_3(x_2, x_1, x_0, y_2, y_1, y_0) = x_2 \bar{y}_2 + x_2 y_2 x_1 \bar{y}_1 + \bar{x}_2 \bar{y}_2 x_1 \bar{y}_1$$

$$+ x_2 y_2 x_1 y_1 x_0 \bar{y}_0$$

$$+ x_2 y_2 \bar{x}_1 \bar{y}_1 x_0 \bar{y}_0 + \bar{x}_2 \bar{y}_2 x_1 y_1 x_0 \bar{y}_0$$

$$+ \bar{x}_2 \bar{y}_2 \bar{x}_1 \bar{y}_1 x_0 \bar{y}_0$$

$$= Z_1(x_2, y_2) + Z_1(x_1, y_1) E_1(x_2, y_2)$$

$$+ Z_1(x_0, y_0)(E_1(x_2, y_2) E_1(x_1, y_1))$$

$$= Z_2(x_2, x_1, y_2, y_1) + Z_1(x_0, y_0) E_2(x_2, x_1, y_2, y_1)$$

where $\qquad E_2(x_2, x_1, y_2, y_1) = E_1(x_2, y_2) \cdot E_1(x_1, y_1).$

In general, a $k$-bit comparison circuit can be defined as

$$Z_k(x_{k-1}, \ldots, x_1, x_0, y_{k-1}, \ldots, y_1, y_0) = Z_{k-1}(x_{k-1}, \ldots, x_1,$$

$$y_{k-1}, \ldots, y_1) + Z_1(x_0, y_0) E_{k-1}(x_{k-1}, \ldots, x_1, y_{k-1}, \ldots, y_1)$$

where $\qquad E_{k-1}(x_{k-1}, \ldots, x_1, y_{k-1}, \ldots, y_1)$

$$= E_1(x_{k-1}, y_{k-1}) \cdot E_1(x_{k-2}, y_{k-2}) \ldots \cdot E_1(x_1, y_1)$$

Assuming $Z_{k-1}$ and $E_{k-1}$ are inputs we can design the simple circuit of Figure 4.10(a) to realize $Z_k$. If we design this circuit so that it also generates the output $E_k(x_{k-1}, \ldots, x_1, x_0, y_{k-1}, \ldots, y_1, y_0) = E_{k-1}$ $(x_{k-1}, \ldots, x_1, y_{k-1}, \ldots, y_1)(x_0 y_0 + \bar{x}_0 \bar{y}_0)$ (Figure 4.10(b)) a cascade of $k$ such circuits is a $k$-bit comparison circuit where $Z_0 = 0$, $E_0 = 1$ (Figure 4.11). We have thus generated a multiple level $k$-bit comparison circuit. In addition to requiring significantly fewer gates and gate inputs than the equivalent two level circuit, this realization also has the advantage

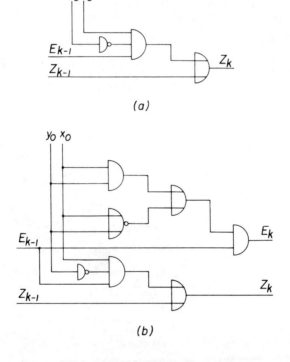

(a)

(b)

Figure 4.10    One-bit comparator module

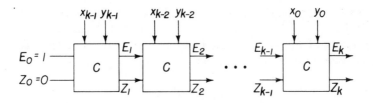

Figure 4.11   Modular design of comparator

of being *modular*. That is, it is composed of identical subcircuits, called *cells* or *modules*, arranged in the form of a *linear cascade*. This simplifies the construction of the circuit and also enables us to convert it to a $(k + 1)$-bit comparison circuit simply by adding another cell to the $k$-bit circuit. However, the many level circuit is considerably slower than the two level circuit. If we assume each gate has a delay $\Delta$ associated with it, then each cell (Figure 4.10(b)) has a delay of $\Delta' = 2\Delta$ and the cascade of $k$ cells has a delay of $k\Delta'$. The corresponding two level realization only has delay $2\Delta$.

The derivation of the circuit of Figure 4.11 required factoring in a particular way which required some ingenuity. However, the same circuit could have been derived in a much simpler way. Let us consider the procedure by which two numbers would actually be compared. If

$$\mathbf{X} = 1\,0\,0\,1\,1\,0\,1\,0\,1\,1\,1\,0\,0\,1\,0\,1$$

and

$$\mathbf{Y} = 1\,0\,0\,0\,1\,0\,1\,1\,0\,1\,0\,1\,0\,1\,0\,0$$

we could compare these numbers and determine that $\mathbf{X} > \mathbf{Y}$ without determining what the corresponding numbers were in decimal. Effectively, we would scan the two numbers from left to right (i.e., most significant bit first) until we reached a position in which they were unequal. If in that position $y_i > x_i$ then $\mathbf{Y} > \mathbf{X}$, and if $x_i > y_i$ then $\mathbf{X} > \mathbf{Y}$, independent of the remaining less significant bits of $\mathbf{X}$ and $\mathbf{Y}$. We will now design a circuit which in effect implements this algorithm. When examining bit $i$ we will generate a 1 output if $x_i = 1$, $y_i = 0$ and in all previous bits $j > i$, $y_j = x_j$, or if we can conclude as a result of examining bits $j$, $j > i$, that $\mathbf{X} > \mathbf{Y}$. Therefore, the $i^{\text{th}}$ bit cell has two inputs $E_{i-1}$, which has the value 1 if and only if $y_j = x_j$ for all $j > i$, and $Z_{i-1}$, which has the value 1 if and only if $X > Y$ independent of the value of bit $i$ and all less significant bits. The basic cell has two outputs.

$$Z_i = Z_{i-1} + E_{i-1} \cdot x_i \bar{y}_i$$
$$E_i = E_{i-1} \cdot (x_i y_i + \bar{x}_i \bar{y}_i)$$

which is the cell of Figure 4.10(b) and the $k$-bit comparator is realized by a cascade of $k$ such cells.

## 4.2.2 PARALLEL ADDER

The concept of designing a circuit to implement an algorithm can also be used to design a *parallel adder*. This circuit will be designed to implement the binary addition algorithm of Procedure 2.3. The addition of two $n$-bit numbers $X = x_{n-1} x_{n-2} \dots x_1 x_0$ and $Y = y_{n-1} \dots y_1 y_0$ is performed by adding the numbers bit by bit proceeding from the least to the most significant bits (right to left). The $i^{\text{th}}$ bit of the sum $Z_i$ is generated by adding $x_i, y_i$ and the carry, $C_i$, from the $(i-1)^{\text{st}}$ bit addition. Thus a circuit to perform the addition of bit $i$ would have three inputs $(x_i, y_i, C_i)$ and generate two outputs $(Z_i, C_{i+1})$, as specified by the truth table of Figure 4.12(a) and realized by the circuit of Figure 4.12(b). Connecting $n$ such circuits results in an $n$-bit parallel adder (Figure 4.13) where $C_0 = 0$.

The parallel adder* requires much less logic than an adder designed as a two level combinational circuit would require, but it is much slower, again demonstrating the possible tradeoff between speed and hardware requirements. However, these are the two extreme cases in this tradeoff. It is possible to design adders which are faster than the parallel adder and require less logic than the two level combinational circuit adder. One such circuit can be designed as a cascade of $n/2$ modules, each of which is a 2-bit adder designed as a two level combinational circuit. The basic module has 5 inputs $(x_{i-1}, x_i, y_{i-1}, y_i, C_{i-1})$ and 3 outputs $(Z_{i-1}, Z_i, C_{i+1})$. The Karnaugh maps describing this module are shown

---

*The parallel adder has been called a *ripple carry* adder since the effect of a carry from module $i$ will propagate (ripple) to cell $j$ if all intermediate cells have the $x$ and/or $y$ input equal to 1.

| $x_i$ | $y_i$ | $C_i$ | $Z_i$ | $C_{i+1}$ |
|-------|-------|-------|-------|-----------|
| 0 | 0 | 0 | 0 | 0 |
| 0 | 0 | 1 | 1 | 0 |
| 0 | 1 | 0 | 1 | 0 |
| 0 | 1 | 1 | 0 | 1 |
| 1 | 0 | 0 | 1 | 0 |
| 1 | 0 | 1 | 0 | 1 |
| 1 | 1 | 0 | 0 | 1 |
| 1 | 1 | 1 | 1 | 1 |

(a)

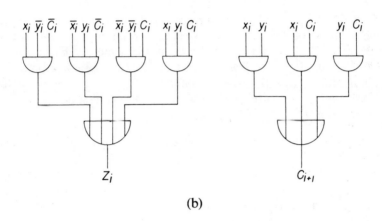

(b)

**Figure 4.12   One-bit adder module**

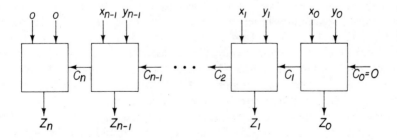

**Figure 4.13   Modular design of parallel adder**

in Figure 4.14. The minimal two level multiple output realization of this basic cell has cost in excess of twice the cost of the basic cell of the parallel adder (See Problem 4.6). However, an $n$-bit adder designed from such 2-bit adder modules will have $n$ levels ($n/2$ two level modules) of logic compared with $2n$ levels for the parallel adder assuming the modules are two level circuits in both cases. Thus this circuit is approximately twice as fast as the parallel adder. This design concept can be generalized to design an $n$-bit adder as a cascade of $k$ $n/k$-bit two level combinational circuit adders. The two level $n$-bit adder and the parallel adder represent the extreme cases $k = 1$ and $k = n$ respectively.

Let us consider a $3n$-bit adder circuit designed as a cascade of $n$ two level 3-bit adders. A typical cell has 7 inputs $(x_i, y_i, x_{i-1}, y_{i-1}, x_{i-2}, y_{i-2}, C_{i-2})$ and 4 outputs $(Z_i, Z_{i-1}, Z_{i-2}, C_{i+1})$. The speed of the adder $A$ depends on the number of levels of $A$ which depends on the number of levels in the 3-bit adder cell of the output $C_{i+1}$. By realizing $C_{i+1}$ as a two level circuit the adder $A$ has $2n$ levels, and is thus about three times as fast as the comparable ripple carry (parallel) adder. Note that the outputs $(Z_i, Z_{i-1}, Z_{i-2})$ can be realized as multilevel circuits in the same manner as the parallel adder.

The function $C_{i+1}$ can be expressed as

$$C_{i+1} = T_{i+1} + T_i S_{i+1} + T_{i-1} S_i S_{i+1} + C_{i-2} S_{i-1} S_i S_{i+1}$$

where

$$S_i = x_i \bar{y}_i + \bar{x}_i y_i$$

and

$$T_i = x_i y_i.$$

If realized as a two level circuit, this requires about 16 gates with 105 gate inputs, excluding inverters, and maximum fan-in of 15. However, if $C_{i+1}$ is realized as a four level circuit, as shown in Figure 4.15, the total amount of logic required is reduced to 14 gates with 35 gate inputs and maximum fan-in of 4, and the total number of levels of $A$ is $2n + 2$ since the number of levels from $C_{i-2}$ to $C_{i+1}$ in each cell is two. This circuit has been referred to as a *carry lookahead adder*[12].

Modular realizations can frequently be derived for functions specified on sets of related input bits, which can be thought of as *words*, if the function effectively requires the same computation on each of the bits.

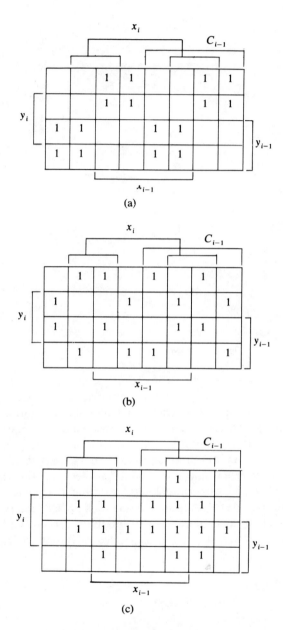

Figure 4.14  (a) Map of $Z_{i-1}$, (b) Map of $Z_i$, (c) Map of $C_{i+1}$

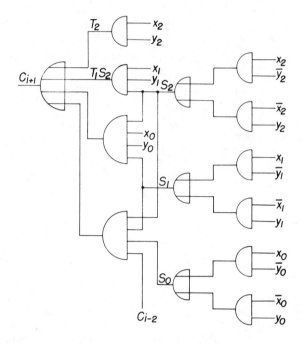

**Figure 4.15    Module of carry lookahead adder**

Modular realizations which consist of linear cascades of identical cells of combinational logic have been referred to as *1-dimensional iterative arrays*[9]. For functions which have such modular realizations, the general speed/complexity tradeoff can often be made more dramatic by realizing such functions as *sequential circuits.* In these circuits, which we shall consider in Chapter 5, bit operations can be performed sequentially rather than in parallel, and one module is used in a time sharing mode to generate all such bit operations thus resulting in a significant reduction in hardware cost. In effect the spatial relationship between bits at different positions *i* and *j* of a linear cascade of modules, is replaced by a temporal relationship affecting a single bit at two different instants of time *i* and *j.*

## 4.3 LOGICAL COMPLETENESS

In Chapter 3 we considered the problem of designing combinational circuits using very simple elements called gates. With the advent of integrated circuits and especially large scale integration (LSI) it becomes feasible to use much more complex circuits as basic elements. In Chapter 6 we shall show how such complex modules can be interconnected to form systems. In the previous section of this chapter we have seen how some functions can be designed by interconnecting identical modules. A fundamental question arises in connection with the design of functions from non-elementary modules. Given a specific collection of different modules $S$ corresponding to a set of combinational functions $\{f_1, f_2 \ldots f_p\}$, can *all* logic functions be designed by interconnecting modules of the basic set $S$?

In the previous chapters we have seen that all combinational functions can be realized using only 2-input AND and 2-input OR gates assuming double rail inputs, or from 2-input AND, 2-input OR, and NOT gates assuming single rail inputs. Furthermore the 2-input NAND gate can be used to realize all functions, as can the 2-input NOR gate. Thus the function sets $\{f_1 = \overline{x_1 + x_2}\}$, $\{f_2 = \overline{x_1 \cdot x_2}\}$ are logically (functionally) complete in this sense. A set of functions $S = \{f_1, f_2 \ldots f_p\}$ is *strong logically complete* if all combinational functions (including 0 and 1) can be realized from functions of this set using only the single rail input variables $x_i$ as inputs (i.e., the constant inputs 0 and 1 are not used). The set is *weak logically complete* if it is not strong complete but all combinational functions can be realized using the single rail input variables $x_i$ *and* the constant inputs 0 and 1.*

It is sometimes easy to demonstrate that a given set $S$ is strong complete. To do so we must only show how the element functions of $S$ can be combined to realize all functions of some set $S'$ which is known to be strong complete. $S'$ can be the NAND function, the NOR function, or any other function set which has previously been determined to be strong complete.

For instance if $S$ consists of the single function $f(x_1, x_2, x_3) = (\bar{x}_1 \bar{x}_2 + x_1 x_3)\bar{x}_4$, we can prove that $S$ is strong logically complete by realizing the NAND element as a composition of $f$ elements as follows. We first notice that $f(x_1, x_1, x_1, x_1) = \bar{x}_1$ so that we can invert any input variable. Using inverted inputs we can produce the AND function as $f(x_1, x_2, x_2, \bar{x}_1) = (\bar{x}_1 \bar{x}_2 + x_1 x_2)\bar{\bar{x}}_1 = x_1 x_2$. We can therefore produce the NAND by $f(z, z, z, z) = \bar{z} = \overline{x_1 x_2}$ if $z = x_1 x_2$. Thus the NAND element is realized by the circuit of Figure 4.16.

---

*We can also define *strong c-complete* and *weak c-complete* where the input variables $x_i$ are double rail. However precise characterizing properties for such sets of functions are not known (See Problem 4.15)

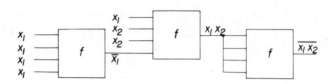

**Figure 4.16    Realization of a NAND gate**

This would *prove* that $S$ is strong logically complete. Of course, the efficient realization of an arbitrary function $f'$ from the basic functions of $S$ may require considerable ingenuity (as may the realization of a NAND from element functions of $S$). To show that a set $S$ is not strong complete it is necessary to *prove* that some function $f'$ cannot be realized from elements of $S$. By studying some basic properties of combinational functions it is possible to derive easily applied necessary and sufficient conditions on logical completeness which greatly facilitate both of these processes. The following five properties must be considered.

**(1) Positiveness**

Consider two points, $A$ and $B$, in an $n$-dimensional space. These· points can be represented as vectors $A = (a_1, a_2 ... a_n)$ and $B = (b_1, b_2 ... b_n)$ where $a_i, b_i = 0$ or $1$ for all $i = 1, 2, ..., n$. If $a_i \leq b_i$ for all $i$ (i.e., if $a_i = 1$, then $b_i = 1$) then $A \leq B$. A function $f(x_1, x_2, ..., x_n)$ is *positive* if and only if $f(A) \leq f(B)$ for all $A \leq B$. A function $F$ can be shown to be positive if and only if a minimal sum of products expression for $F$ has no complemented variables. (See Problem 4.16)

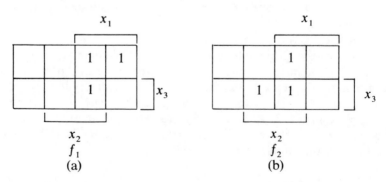

**Figure 4.17    Maps of functions $f_1$ and $f_2$**

**Example 4.2**

The function $f_1(x_1,x_2,x_3)$ of Figure 4.17(a) is not positive since $A$ = (1, 0, 0) $\leq$ $B$ = (1, 0, 1) and $f(1, 0, 0)$ = 1 $\not\leq$ $f(1, 0, 1)$ = 0. The function $f_2(x_1,x_2,x_3) = x_1 x_2 + x_2 x_3$ is positive.

**(2) Zero-preservation**

A function $f(x_1,x_2...x_n)$ is *zero-preserving* if $f(0,0, ...,0) = 0$.

**(3) One-preservation**

A function $f(x_1,x_2...x_n)$ is *one-preserving* if $f(1,1, ...,1) = 1$.

**(4) Self-duality**

A function $f(x_1,x_2,...,x_n)$ is self-dual if the dual of $f$, $f_d$ = $\bar{f}(\bar{x}_1,\bar{x}_2,...,\bar{x}_n)$, is equal to $f(x_1,x_2,...,x_n)$.

**Example 4.3**

The function $f(x_1,x_2,x_3)$ of Figure 4.18(a) is not 0-preserving, and not 1-preserving. To determine self-duality we construct the function $f(\bar{x}_1,\bar{x}_2,\bar{x}_3)$ (Figure 4.18(b)) by interchanging columns (2 and 4) and (1 and 3) and rows (1 and 2) simultaneously. Since $\bar{f}(x_1,x_2,x_3)$ = $f(\bar{x}_1,\bar{x}_2,\bar{x}_3)$, $f$ is self-dual.

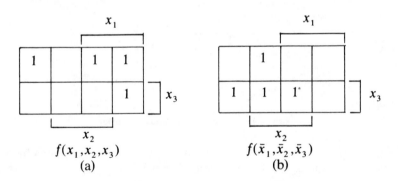

Figure 4.18   Map of $f(x_1,x_2,x_3)$ and $f(\bar{x}_1,\bar{x}_2,\bar{x}_3)$

**(5) Linearity**

Any function can be expressed as a modulo 2 (exclusive OR) sum of products. This expression which is referred to as the Reed-Muller expansion is of the following form

$$f(x_1, x_2 ... x_n) = a_o \oplus a_1 x_1 + a_2 x_2 \oplus \quad \cdots \quad \oplus a_n x_n$$
$$\oplus a_{n+1} x_1 x_2 \oplus a_{n+2} x_1 x_3 \oplus \cdots$$
$$\oplus a_{2^n-1} x_1 x_2 ... x_n \qquad \text{where } a_i = 0 \text{ or } 1$$

which has one product term for each possible subset of the input variables, $2^n$ terms in all (See Problem 3.23). A function $f(x_1, x_2 ... x_n)$ is linear if in its Reed-Muller expansion, $a_i = 0$ for all $i > n$ (i.e., $f(x_1, x_2 ... x_n)$ $= a_o \oplus a_1 x_1 \oplus ... \oplus a_n x_n$, $a_i = 0, 1$). This expression can be obtained for a given function $F$ by expressing $f$ as a ring sum of minterms (or other product terms), replacing each complemented variable $\bar{x}_i$ by $1$ $\oplus x_i$, expanding, and simplifying by $g \oplus g = 0$ for any expression $g$.

**Example 4.4**

The function $f(x_1, x_2, x_3)$ of Figure 4.19 can be expressed as the ring sum of products $\bar{x}_1 \bar{x}_3 \oplus x_1 x_3$. But $\bar{x}_1 \bar{x}_3 = (1 \oplus x_1)(1 \oplus x_3) = 1 \oplus x_1$ $\oplus x_3 \oplus x_1 x_3$. Thus $f = 1 \oplus x_1 \oplus x_3 \oplus x_1 x_3 \oplus x_1 x_3 = 1 \oplus x_1 \oplus x_3$, and hence $f$ is linear.

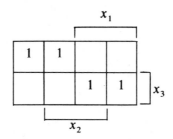

Figure 4.19    Map of combinational function

The set of all combinational functions contains some functions having and some functions not having each of these five properties. We shall now prove that the set $S$ of basic functions must have some function which does not have each of these properties in order for $S$ to be strong complete.

**Theorem 4.1**

Let $S = \{f_1, ..., f_k\}$ be a set of basic element functions all of which are

(a) positive            or
(b) 0-preserving       or
(c) 1-preserving       or
(d) self-dual          or
(e) linear.

Then any combinational function which can be realized by interconnecting these functions will also have the corresponding property, and therefore $S$ is not strong complete.

**Proof**

The proof will be illustrated for the property of 0-preservation. The proofs of the remaining parts of the theorem are similar and will be left as an exercise. Assume $S = \{f_1, f_2, ..., f_k\}$ and each of these functions are 0-preserving. Suppose they are connected in a $k$-level circuit $C$ to realize a function $f'$. Consider all first level elements in $C$ (function blocks whose inputs are input variables $x_i$). The output of all such blocks if $x_1 = x_2 = ... = x_n = 0$ is also 0 because the functions $f_k$ are 0-preserving. Thus the inputs of second level elements of $C$ (whose inputs are input variables or outputs of first level elements) are all 0 if all $x_i$ are 0 and the outputs of these elements are 0. Repeating this argument $k$ times we conclude the output of $C$ is 0 if $x_1 = x_2 = ... = x_n = 0$ and hence $f'$ is 0-preserving.

The constant functions $f(x_1, x_2...x_n) = 0$ and $f(x_1, x_2...x_n) = 1$ are non 1-preserving and non 0-preserving respectively and also non-self-dual. In testing a set of functions $S$ for weak logical completeness these constant functions can be assumed available as inputs and hence $S$ may not require functions which are not 0-preserving and 1-preserving. In fact, we shall show that $S$ must only contain functions which are not positive, and not linear.

**Lemma 4.1**

If the constants 0 and 1 can be used as inputs, any non-positive function $f$ can be made to realize an inverter.

**Proof**

Suppose $f(x_1, x_2...x_n)$ is a non-positive function. Then there exists two points in the $n$-dimensional space $A = (a_1, a_2...a_n)$ and $B = (b_1, b_2...b_n)$ such that $A < B$ and $f(A) = 1$, $F(B) = 0$. Suppose $A$ and $B$ differ in $k \leq n$ variables. Then there are a set of points $A_1, A_2, ..., A_{k-1}$

such that $A_1$ differs from $A$ in one variable, $A_{k-1}$ differs from $B$ in one variable, and $A_i$ differs from $A_{i+1}$ in one variable for all $i$, $1 \le i \le k - 2$. Let $A = A_0$ and $B = A_n$. Since $f(A) = 1$ and $f(B) = 0$, for some $i$ $f(A_i) = 1$ and $f(A_{i+1}) = 0$. If $A_i$ and $A_{i+1}$ differ only in variable $x_j$ and $A_i = (a_{i1}, a_{i2}, ..., a_{in})$. Then $f(a_{i1}, a_{i2}, ..., x_j, a_{i(j+1)}, ..., a_{in}) = \bar{x}_j$.

### Example 4.5

Let $f = x_1 x_2 + \bar{x}_1 x_3 \bar{x}_4$. If $A = (0,0,1,0)$ and $B = (1,0,1,1)$ then $A < B$ and $f(A) = 1$ and $f(B) = 0$. Let $A_1 = (1,0,1,0)$. Then $f(A_1) = 0$ and $f(A) = 1$ and $A_1$ and $A$ differ only in variable $x_1$. Therefore $f(x_1,0,1,0) = \bar{x}_1$.

### Lemma 4.2

If the constant inputs 0 and 1 and inverters can be used, any nonlinear function $f$ can be made to realize AND and OR gates.

### Proof

Suppose $f$ is nonlinear. Then the Reed-Muller expansion of $f$ must contain at least one term containing two or more variables. Suppose $T$ is such a term with the fewest number of variables and let $T$ be the product of $x_i, x_j$ and a set of variables $P$ (which may be the null set). Assign all variables in $P$ the value 1, all other variables except $x_i$ and $x_j$ the value 0. The resulting function is of the form $a_0 \oplus a_1 x_i \oplus a_2 x_j \oplus x_i x_j$. The following table shows the function obtained for all possible values of $a_0, a_1, a_2$. By DeMorgan's laws, each of these functions can be converted to $x_i x_j$ or $x_i + x_j$ by complementing inputs

| $a_0$ | $a_1$ | $a_2$ | Function |
|-------|-------|-------|----------|
| 0 | 0 | 0 | $x_i x_j$ |
| 0 | 0 | 1 | $\bar{x}_i x_j$ |
| 0 | 1 | 0 | $x_i \bar{x}_j$ |
| 0 | 1 | 1 | $x_i + x_j$ |
| 1 | 0 | 0 | $\bar{x}_i + \bar{x}_j$ |
| 1 | 0 | 1 | $x_i + \bar{x}_j$ |
| 1 | 1 | 0 | $\bar{x}_i + x_j$ |
| 1 | 1 | 1 | $\bar{x}_i \bar{x}_j$ |

or outputs. Thus with 0 and 1 inputs and NOT gates, any nonlinear function can be made to realize AND and OR gates.

### Theorem 4.2

A set of logic functions $S$ is weak complete if and only if $S$ contains a function which is non-positive and a function which is nonlinear.

### Proof

Necessity follows directly from Theorem 4.1. Sufficiency follows directly from Lemmas 4.1 and 4.2.

### Theorem 4.3

A set of functions $S = \{f_1,...,f_k\}$ is strong complete if and only if $S$ contains at least one function with each of the following properties: (a) non-positive, (b) non 0-preserving, (c) non 1-preserving, (d) non self-dual, (e) nonlinear. (For each of these properties, a different function of $S$ may have that property).

### Proof

Necessity follows directly from Theorem 4.1. To prove sufficiency we will show how $S$ can be used to realize the constants 0 and 1. The result will then follow directly from Theorem 4.2. There are two cases which must be considered.

**Case 1:** There exists a function $f$ in $S$ which is both non 0-preserving and non 1-preserving. That is, $f(0,0,...,0) = 1$ and $f(1,1,...,1) = 0$. Then $f(x_1,x_1,...,x_1) = \bar{x}_1$. Now let $f_2$ be any non self-dual function. Then there exist two points $A = (a_1,a_2,...,a_n)$ and $B = (\bar{a}_1,\bar{a}_2,...,\bar{a}_n)$ such that $f_2(A) = f_2(B)$. The outputs 0 and 1 can be realized by the circuit of Figure 4.20.

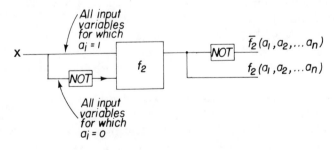

Figure 4.20    Realization of 0 and 1 for case 1.

**Case 2:** Assume there is no function in $S$ which is both non 0-preserving and non 1-preserving. Then there exist two functions $f_0$ and $f_1$ such that $f_0(0,0, ...,0) = 0 = f_0(1,1, ...,1)$ and $f_1(0,0, ...,0) = 1 = f_1(1,1, ...,1)$. Then $f_0(x_1,x_1, ...,x_1) = 0$ and $f_1(x_1, ...,x_1) = 1$.

**Example 4.6**

(a) Let $S = \{f\}$ where $f = \bar{x}_1 \bar{x}_2 + x_1 x_3$. Then $S$ has no function which is non 1-preserving and hence $S$ is not strong complete.

(b) Let $S = \{f'\}$ where $f' = (\bar{x}_1 \bar{x}_2 + x_1 x_3)\bar{x}_4$. Since $f'(0,0,0,0) = 1$ and $f'(1,1,1,1) = 0$, $f'$ is non 0-preserving and non 1-preserving. Furthermore $f'$ is non-positive, non self-dual (since $f'(0,1,0,0) = f'(1,0,1,1) = 0$) and non-linear since $f' = 1 \oplus x_1 \oplus x_2 \oplus x_4 \oplus x_1 x_2 \oplus x_1 x_3 \oplus x_1 x_4 \oplus x_2 x_4 \oplus x_1 x_2 x_4 \oplus x_1 x_3 x_4)$. Thus $S$ is strong complete. The function $f'$ can be used to realize the constants 0 and 1 by the circuit of Figure 4.21.

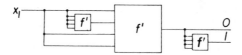

Figure 4.21   Realization of 0 and 1 using $f'$.

A strong (weak) complete set of logic functions $S$ is a *strong (weak) basis* if no proper subset of $S$ is *strong (weak complete)*. For every non 0-preserving function $f$, $f$ is also non 1-preserving or non self-dual (see Problem 4.7). However, for all other properties of Theorem 4.3 there exist functions having only one of these properties. Hence the maximum number of functions in a strong basis is four. The minimum number of functions in a strong (or weak) basis is one.

Since there exist functions which are nonlinear but are positive (e.g., AND) and non-positive but linear (e.g., $x_1 \bar{x}_2 + \bar{x}_1 x_2$), from Theorem 4.2 it follows that the maximum number of functions in a weak basis is two.

It is also of interest to consider the relative efficiency of different sets of logically complete modules in realizing arbitrary functions. It seems especially difficult to use a module with output $f$ to realize the dual of $f$, $f_d$. Thus it requires five 3-input NAND gates to realize a 3-input NOR. For this reason it seems likely that sets of primitives which contain dual pairs may be relatively efficient. Thus the complete set {NAND, AND} may not be as good as the set {NAND, NOR}. However this hypothesis has not been verified and little has been achieved in measuring efficiency or obtaining efficient realizations using sets of complex modules.

## 4.4 CLASSIFICATIONS OF COMBINATIONAL FUNCTIONS

Combinational functions have been classified according to different properties. Frequently these properties are of significance primarily for one particular technology.

### 4.4.1 SYMMETRIC FUNCTIONS

An interesting class of combinational functions are *symmetric functions*. A function $f(x_1, x_2, ..., x_n)$ is *symmetric in variables* $x_i, x_j$, denoted by $x_i \sim x_j$, if $f(x_1, x_2, ..., x_i, ..., x_j, ..., x_n) = f(x_1, x_2, ..., x_j, ..., x_i, ..., x_n)$. That is, interchanging variables $x_i$ and $x_j$ on a Karnaugh map of $f$ results in an identical map, or equivalently replacing all appearances of $x_i$ by $x_j$ and $x_j$ by $x_i$ in a Boolean expression for $f$ results in an equivalent expression. Similarly, $x_i \sim \bar{x}_j$ if $f(x_1, x_2, ..., x_i, ..., x_j, ..., x_n) = f(x_1, x_2, ..., \bar{x}_j, ..., \bar{x}_i, ..., x_n)$. Consider the function $f(x_1, x_2, x_3) = \bar{x}_2 + \bar{x}_1 \bar{x}_3 + x_1 x_3$. Interchanging $x_1$ and $x_3$ in this expression results in $f(x_3, x_2, x_1) = \bar{x}_2 + \bar{x}_3 \bar{x}_1 + x_3 x_1 = f(x_1, x_2, x_3)$. Therefore $x_1 \sim x_3$. However $f(x_2, x_1, x_3) = \bar{x}_1 + \bar{x}_2 \bar{x}_3 + x_2 x_3 \neq f(x_1, x_2, x_3)$, and therefore $x_1 \not\sim x_2$. The symmetry relation is transitive, i.e. if $x_i \sim x_j$ and $x_j \sim x_k$ then $x_i \sim x_k$ (Exercise). A function $f(x_1, x_2, ..., x_n)$ is *totally symmetric* if for every pair of variables $x_i, x_j, x_i \sim x_j$ and/or $x_i \sim \bar{x}_j$. For the previously considered function, $x_1 \not\sim x_2$ and $x_1 \not\sim \bar{x}_2$. Therefore $f$ is *not* totally symmetric. Define the function $S_i(x_1^*, x_2^*, ..., x_n^*)$ which is 1 if and only if exactly $i$ of the $n$ variables are 1, where $x_i^*$ is either $x_i$ or $\bar{x}_i$. (See Problem 3.17). Similarly $S_{i_1, i_2, ..., i_p}(x_1^*, x_2^*, ..., x_n^*)$ is 1 if and only if exactly $i_1$ or $i_2$ or ... or $i_p$ of the $n$ variables are 1. The class of all functions of this type is identical to the class of totally symmetric functions (Problem 4.14).

**Determination of (Total) Symmetry**

Using Shannon's expansion theorem, any combinational function of

$n$ variables may be expressed as:

$$f(x_1, x_2, ...,x_n) = x_1 f(1,x_2,x_3, ...,x_n) + \bar{x}_1 f(0,x_2,x_3, ...,x_n) \qquad (1)$$

$$= x_2 f(x_1,1,x_3, ...,x_n) + \bar{x}_2 f(x_1,0,x_3, ...,x_n) \qquad (2)$$

If $x_1 \sim x_2$ then (1) $= x_2 f(1,x_1,x_3, ...,x_n) + \bar{x}_2 f(0,x_1,x_3, ...,x_n) \qquad (3).$

Since (2) = (3), $f(x_1,1,x_3, ...,x_n) = f(1,x_1,x_3, ...,x_n)$ and $f(x_1,0,x_3, ...,x_n) = f(0,x_1, x_3, ...,x_n)$. If $f(x_1,1,x_3, ...x_n) = f(1,x_1,x_3, ...,x_n)$, then the number of 1-points in the two functions must be equal. Similarly if $x_1 \sim \bar{x}_2$, then $f(1,x_2,x_3, ...,x_n)$ and $f(x_1,0,x_3, ...,x_n)$ must have the same number of 1-points.

Thus given a function $f(x_1,x_2, ...,x_n)$, if $f$ has $p_i$ 1-points with $x_i = 1$ and $q_i$ 1-points with $x_i = 0$ then $f$ is totally symmetric *only if* for all $i$ and $j$ either $p_i = p_j$ and $q_i = q_j$ (for $x_i \sim x_j$) or $p_i = q_j$ and $p_j = q_i$ (for $x_i \sim \bar{x}_j$).

This condition (which is called the *ratio test*) is necessary but not sufficient for symmetry. However it is easy to apply and frequently facilitates the determination of symmetry by limiting the possible variables of symmetry.

### Example 4.7

Let $f(x_1,x_2,x_3,x_4)$ be a function whose 1-points are listed in the following table. If $p_i$ is the number of 1-points with $x_i = 1$ and $q_i$

| $x_1$ | $x_2$ | $x_3$ | $x_4$ |
|---|---|---|---|
| 0 | 0 | 0 | 1 |
| 0 | 0 | 1 | 0 |
| 0 | 1 | 1 | 1 |
| 1 | 0 | 0 | 0 |
| 1 | 0 | 0 | 1 |
| 1 | 1 | 0 | 1 |
| 1 | 1 | 1 | 0 |

is the number of 1-points with $x_i = 0$ then

$$p_1 = 4 \quad p_2 = 3 \quad p_3 = 3 \quad p_4 = 4$$
$$q_1 = 3 \quad q_2 = 4 \quad q_3 = 4 \quad q_4 = 3$$

We can conclude that $x_1$ may be symmetric to $\bar{x}_2$ (since $p_1 = q_1$, but $x_1 \not\sim x_2$ since $p_1 \neq p_2$). Similarly $x_1$ may be symmetric to $\bar{x}_3$ and $x_4$. Thus if $f$ is symmetric it must be of the form $S_{i_1,\ldots,i_p}(x_1,\bar{x}_2,\bar{x}_3,x_4)$. Row 1 of the list of 1-points has three $(\bar{x}_2,\bar{x}_3,x_4)$ of those four variables equal to 1. However the point (1 0 1 1) which also has three $(x_1,\bar{x}_2,x_4)$ of these four variables equal to 1 is not a 1-point. Therefore $f$ is not symmetric.

The following theorem can be used to show symmetry for a given function.

**Theorem 4.4**

For a function $f(x_1,x_2,\ldots,x_n)$: (a) $x_1 \sim x_2$ if and only if $f(0,1,x_3, x_4,\ldots,x_n) = f(1,0,x_3,x_4,\ldots,x_n)$ and (b) $x_1 \sim \bar{x}_2$ if and only if $f(0,0,x_3,x_4,\ldots,x_n) = f(1,1,x_3,x_4,\ldots,x_n)$.

**Proof**

(a) By the Shannon expansion theorem $f(x_1,x_2,\ldots,x_n) = x_1 x_2 f(1,1,x_3, \ldots,x_n) + x_1 \bar{x}_2 f(1,0,x_3,\ldots,x_n) + \bar{x}_1 x_2 f(0,1,x_3,\ldots,x_n) + \bar{x}_1 \bar{x}_2 f(0,0,x_3, \ldots,x_n)$.

If $x_1 \sim x_2$ then we can interchange $x_1$ and $x_2$ without changing the function. If we do this to the right side of the above equation we obtain the expression $x_1 x_2 f(1,1,x_3,\ldots,x_n) + \bar{x}_1 x_2(1,0,x_3,\ldots,x_n) + x_1 \bar{x}_2 f(0,1,x_3,\ldots,x_n) + \bar{x}_1 \bar{x}_2 f(0,0,x_3,\ldots,x_n)$. This expression is equal to the original expression from which it was obtained if and only if $f(0,1,x_3, \ldots,x_n) = f(1,0,x_3,\ldots,x_n)$.

(b) Exercise

In conjunction with the ratio test, this result can be used to determine symmetric functions.

**Example 4.8**

Consider the function whose 1-points are listed in the following table. By the ratio test, $x_1$ may be symmetric to $\bar{x}_2,\bar{x}_3,x_4$. Applying Theorem 4.4 to test for $x_1 \sim \bar{x}_2$, we find that the 1-points of both the subfunctions $f(0,0,x_3,x_4)$ and $f(1,1,x_3,x_4)$ are $x_3 = x_4 = 0$ and $x_3 = x_4 = 1$. Therefore $x_1 \sim \bar{x}_2$. Similarly it is easily shown that $x_1 \sim \bar{x}_3$ and $x_1 \sim x_4$. Therefore $f = S_{i_1,i_2,\ldots}(x_1,\bar{x}_2,\bar{x}_3,x_4)$. The first six 1-points of the original table have two of the four variables $(x_1,\bar{x}_2,\bar{x}_3,x_4)$ equal to 1, and the last

| $x_1$ | $x_2$ | $x_3$ | $x_4$ |
|-------|-------|-------|-------|
| 1 | 0 | 1 | 0 |
| 1 | 1 | 0 | 0 |
| 1 | 1 | 1 | 1 |
| 0 | 0 | 0 | 0 |
| 0 | 0 | 1 | 1 |
| 0 | 1 | 0 | 1 |
| 1 | 0 | 0 | 0 |
| 1 | 1 | 0 | 1 |
| 1 | 0 | 1 | 1 |
| 0 | 0 | 0 | 1 |

$$p_1 = 6 \qquad p_2 = 4 \qquad p_3 = 4 \qquad p_4 = 6$$

$$q_1 = 4 \qquad q_2 = 6 \qquad q_3 = 6 \qquad q_4 = 4$$

four 1-points have three of these variables equal to 1. Therefore $f = S_{2,3}(x_1, \bar{x}_2, \bar{x}_3, x_4)$.

Harrison[5] has shown that a function $f(x_1, x_2, \ldots, x_n)$ is totally symmetric in the uncomplemented variables, (i.e. $x_1 \sim x_2 \sim x_3 \ldots \sim x_n$) if and only if

$$f(x_1, x_2, \ldots, x_{n-1}, x_n) = f(x_1, x_2, \ldots, x_n, x_{n-1})$$

and
$$f(x_1, x_2, \ldots, x_{n-1}, x_n) = f(x_2, x_3, \ldots, x_{n-1}, x_1 x_n)$$

The proof of this result requires knowledge of mathematical concepts beyond the scope of this book.

### 4.4.2 UNATE FUNCTIONS

Another class of combinational functions are the *unate* functions. Let $x_i$ and $x_j$ be points in the $n$-cube defined by the variables $(x_1, x_2, \ldots, x_n)$. For a function $f(x_1, x_2, \ldots, x_n)$, $f(\mathbf{x}_i) \geq f(\mathbf{x}_j)$ if

$$f(\mathbf{x}_i) = f(\mathbf{x}_j) \text{ or } f(\mathbf{x}_i) = 1 \text{ and } f(\mathbf{x}_j) = 0.$$

The function $f$ is *positive in a variable* $x_i$, if for all $2^{n-1}$ possible combinations of values of the remaining $n-1$ variables.

$$f(x_1,x_2, ...,x_{i-1},1,x_{i+1}, ...,x_n) \geqq f(x_1,x_2, ...,x_{i-1},0,x_{i+1}, ...,x_n)$$

Similarly, $f$ is *negative in the variable* $x_i$ if

$$f(x_1,x_2, ...,x_{i-1},0,x_{i+1}, ...,x_n) \geqq f(x_1,x_2, ...,x_{i-1},1,x_{i+1}, ...,x_n).$$

**Theorem 4.5**

A completely specified combinational function $f(x_1,x_2, ...,x_n)$ is positive in a variable $x_i$ if and only if any minimal sum of products expression for $f$ does not contain the literal $\bar{x}_i$. Similarly, $f$ is negative in the variable $x_i$ if and only if no minimal sum of products expression for $f$ contains the literal $x_i$.

**Proof:** Exercise

A function $f(x_1,x_2, ...,x_n)$ is *positive* if it is positive in all its variables, $x_i$. Similarly, a function which is negative in all its variables is called a *negative* function. A function $f(x_1,x_2, ...,x_n)$ is *unate* if for every $x_i$, $i = 1,2, ...,n$, $f$ is either positive or negative in the variable $x_i$.

**Theorem 4.6**

A function $f(x_1,x_2, ...,x_n)$ is unate if and only if any minimal sum of products expression contains either the literal $x_i$ or $\bar{x}_i$ but not both, for all $x_i$, $i = 1,2, ...,n$.

**Proof:** Exercise.

**Example 4.9**

The function $f_1 = x_1 x_2 + \bar{x}_3 \bar{x}_4$ is unate. However $f_2 = x_1 x_2 x_3 + \bar{x}_1 x_4$ is not unate since both $x_1$ and $\bar{x}_1$ appear in the minimal sum of products expression for $f_2$.

**4.4.3 THRESHOLD FUNCTIONS**

A function $f(x_1,x_2, ...,x_n)$ is a *threshold function* if there exists a set of numbers $w_1,w_2, ...,w_n$ (called *weights*) and a number $T$ (called the *threshold*) such that $f(x_1,x_2, ...,x_n) = 1$ if and only if $\Sigma_{i=1}^{n} w_i x_i \geqq T$ where $x_i = 0$ or $1$ and the multiplication and summation are arithmetic (rather than Boolean). The function $f(x_1,x_2,x_3) = x_1 x_2 + x_2 x_3$ is a threshold function with weights $w_1 = w_3 = 1$, $w_2 = 2$, and threshold $T = 3$, since as shown in the table of Figure 4.22, $f = 1$ if and only if $\Sigma w_i x_i \geq T$.

| $x_1$ | $x_2$ | $x_3$ | $\Sigma w_i \cdot x_i$ | $f$ |
|------|------|------|------|------|
| 0 | 0 | 0 | 0 | 0 |
| 0 | 0 | 1 | 1 | 0 |
| 0 | 1 | 0 | 2 | 0 |
| 0 | 1 | 1 | 3 | 1 |
| 1 | 0 | 0 | 1 | 0 |
| 1 | 0 | 1 | 2 | 0 |
| 1 | 1 | 0 | 3 | 1 |
| 1 | 1 | 1 | 4 | 1 |

Figure 4.22   Tabular description of a threshold function

Threshold functions can be realized using a *threshold element* which is represented as shown in Figure 4.23. However no such element exists in most technologies.

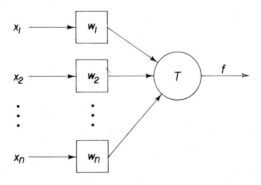

Figure 4.23   A threshold element representation

The following theorem shows that the set of threshold functions is a proper subset of the set of unate functions.

**Theorem 4.7**
(a) All threshold functions are unate.
(b) Not all unate functions are threshold functions.

**Proof**
(a) Assume $f(x_1, x_2, ..., x_n)$ is a threshold function with weights $w_1, w_2,$ $..., w_n$. If $f$ is not unate there is a minimal sum of products expression in which some variable $x_i$ appears complemented and uncomplemented and $f$ can be represented as $f = x_i f_{i1}(x_1, x_2, ..., x_{i-1}, x_{i+1}, ..., x_n) + \bar{x}_i f_{i0}(x_1, x_2, ..., x_{i-1}, x_{i+1}, ..., x_n)$ where $f_{i1} = f(x_1, x_2, ..., x_{i-1}, 1, x_{i+1}, ..., x_n)$ and $f_{i0} = f(x_1, x_2, ..., x_{i-1}, 0, x_{i+1}, ..., x_n)$.
If $w_i \leq 0$, then any 1-point of $f_{i1}$ is also a 1-point of $f_{i0}$. Therefore, $f = \bar{x}_i f_{i0} + f_{i1}$ and $f$ is negative in $x_i$. Contradiction. If $w_i \geq 0$, then any 1-point of $f_{i0}$ is also a 1-point of $f_{i1}$. Therefore, $f = x_i f_{i1} + f_{i0}$ and $f$ is positive in $x_i$. Contradiction. This proves that there is no variable $x_i$ such that both $x_i$ and $\bar{x}_i$ appear in a minimal sum of products expression of $f$. Therefore $f$ is unate.
(b) Consider the unate function $f = x_1 x_2 + x_3 x_4$, and assume that it is a threshold function with weights $w_1, w_2, w_3, w_4$ and threshold $T$. Since $f(1,1,0,0) = 1$, $w_1 + w_2 \geq T$ and since $f(1,0,1,0) = 0$, $w_1 + w_3 < T$. From these inequalities, we can conclude that $w_2 > w_3$. Similarly $f(0,0,1,1) = 1$ implies $w_3 + w_4 \geq T$ and $f(0,1,0,1) = 0$ implies $w_2 + w_4 < T$. From these inequalities, we conclude that $w_3 > w_2$. This contradicts $w_2 < w_3$ and proves that $f$ is not a threshold function.

For a threshold function $f(x_1, x_2, ..., x_n)$, the $2^n$ points of the $n$-cube can be partitioned into two disjoint sets, the 1-points and the 0-points by a hyperplane defined by the equation $w_1 x_1 + w_2 x_2 + ... + w_n x_n = T$. That is all points on one side of this hyperplane (or on it) are 1-points and all points on the other side of the hyperplane are 0-points. Threshold functions are also called *linearly separable* functions, because of the above property.
For each 1-point of $f$, $\mathbf{x}_i = (x_{i1}, x_{i2}, ..., x_{in})$ and each 0-point of $f$, $\mathbf{x}_j = (x_{j1}, x_{j2}, ...x_{jn})$, $\sum_{k=1}^{n} w_k x_{ik} > \sum_{k=1}^{n} w_k x_{jk}$. Thus if $f$ has $k_0$ 0-points and $k_1$ 1-points we can derive a set of $k_0 \cdot k_1$ linear constraints on the weights $w_i$ which can be successfully satisfied if and only if $f$ is a threshold function. It is frequently possible to define a subset of these contraints which if satisfied imply that all constraints are satisfied, by considering those constraints defined by *minimal* 1-points and *maximal* 0-points (changing *any* positive (negative) variable to 0(1) changes a

minimal 1-point to a 0-point, changing *any* positive (negative) variable to 1(0) changes a maximal 0-point to a 1-point).

**Example 4.10**

Consider the unate function $f(x_1,x_2,x_3,x_4) = x_1 x_2 \bar{x}_3 + x_2 \bar{x}_3 \bar{x}_4$. The following constraints (and the conditions $w_1, w_2 > 0$, $w_3, w_4 < 0$) are defined by pairs of points which are minimal 1-points and maximal 0-points.

| 1-point | 0-point | Constraint |
|---------|---------|------------|
| (1 1 0 1) | (1 1 1 0) | $w_1 + w_2 + w_4 > w_1 + w_2 + w_3 \rightarrow w_4 > w_3$ |
| (1 1 0 1) | (1 0 0 0) | $w_1 + w_2 + w_4 > w_1 \rightarrow w_2 + w_4 > 0$ |
| (0 1 0 0) | (1 1 1 0) | $w_2 > w_1 + w_2 + w_3 \rightarrow 0 > w_1 + w_3$ |
| (0 1 0 0) | (1 0 0 0) | $w_2 > w_1$ |

Figure 4.24   Table of constraints for a function

Since $f$ is positive in $x_1, x_2$, then $w_1, w_2 > 0$ and since $f$ is negative in $x_3, x_4$, then $w_3, w_4 < 0$. These constraints are all satisfied by $w_1 = 1$, $w_2 = 2$, $w_3 = -2$, $w_4 = -1$, (which also satisfy all other constraints). The value of $T$ is found from the min $\Sigma w_i x_i$ for all 1-points. In this case $T = 2$.

### 4.4.4 FANOUT-FREE FUNCTIONS

Fanout-free circuits are of significance since they are relatively easy to diagnose [4]. Furthermore in some recent technologies, such as magnetic bubble logic, fanout is difficult to implement [14]. A circuit $N$ is fanout-free if every input and gate output of $N$ is connected to at most one gate-input. A combinational function $f$ is fanout-free if and only if it can be realized by a fanout-free circuit. Using mathematical induction it can be shown that all fanout-free functions are unate (Exercise). Thus unateness is a necessary but, as we shall see, not a sufficient condition for fanout-free functions. For a combinational function $f(x_1, x_2, ..., x_n)$ variables $x_i$ and $x_j$ are *adjacent* if and only if $f(x_1, x_2, ..., a_i, ..., x_j, ..., x_n) = f(x_1, x_2,$

$...,x_i, ...,a_j ...,x_n)$ for some constants $a_i, a_j$, equal to 0 or 1. For the function $f = x_1 x_2 x_3 + x_1 x_2 x_4 + x_3 x_5 + x_4 x_5, x_1$ is adjacent to $x_2$ since $f(0,x_2,x_3,x_4,x_5) = f(x_1,0,x_3,x_4,x_5) = x_3 x_5 + x_4 x_5$. The adjacency relation can be shown to be an equivalence relation (exercise) and hence partitions the set of input variables into disjoint subsets. For the previous function these subsets are $\{x_1,x_2\}$, $\{x_3,x_4\}$, and $\{x_5\}$. The function $f$ can be expressed in factored form in terms of subfunctions of these disjoint variable sets. In this case $f = (x_3 + x_4)(x_1 x_2 + x_5)$ which defines a fanout-free realization. In general the subfunctions of $f$ may have to be evaluated to determine if they are fanout-free in order to determine if $f$ is fanout-free. If no pair of variables of $f$ are adjacent then it can be shown that no fanout-free realization exists. Thus $g = x_1 x_2 + x_2 x_3 + x_1 x_3$ is not a fanout-free function. Thus the basic procedure which consists of evaluating adjacency among variables and factoring the function, must be iterated until the subfunctions are trivially fanout-free, or some subfunction is obtained which is not fanout-free.

**Example 4.11**
Let $f = x_1 x_2 x_3 x_4 + x_1 x_2 x_3 x_5 + x_1 x_2 x_4 x_5$. Then the adjacency relation partitions the set of variables into subsets $\{x_1,x_2\}$, $\{x_3\}$, $\{x_4\}$, $\{x_5\}$ and the function $f$ can be factored as $f = x_1 x_2 \cdot g(x_3,x_4,x_5)$ where $g = x_3 x_4 + x_3 x_5 + x_4 x_5$. The subfunction $g$ has no adjacent variables and hence is not fanout-free and therefore $f$ is not fanout-free.

**SOURCES**

An extensive presentation of factorization techniques can be found in Miller [13]. Systematic decomposition techniques were first developed by Ashenhurst [1] and are presented in a book by Curtis [2]. However these techniques are essentially exhaustive in nature. The techniques for the design of high speed adders is due to MacSorley [12]. McCluskey [9] first noted the correspondence between 1-dimensional iterative arrays and sequential circuits. The results on logical completeness were developed by Post [17] and Klir [8] and are contained in a book edited by Mukhopadahyay [15]. Symmetric functions have been studied by Caldwell [2], McCluskey [10] and Harrison [5] among others, and threshold functions by McNaughton [11] and Paull and McCluskey [16]. The results on fanout free functions are due to Hayes [7].

## REFERENCES

[1] Ashenhurst, R. L., "The Decomposition of Switching Functions," *Proceedings of International Symposium on the Theory of Switching*, April, 1957, Ann. Computation Lab. Harvard Univ., vol. 29, pp. 74-116, 1959.

[2] Caldwell, S. H., "Recognition and Identification of Symmetric Functions," *Trans. AIEE*, pt. II, Commun. Electron., vol. 73, pp. 142-146, May, 1954.

[3] Curtis, H. A., *A New Approach to the Design of Switching Circuits*, D. Van Nostrand Co., Inc., Princeton, N.J., 1962.

[4] Friedman, A. D. and P. R. Menon, *Fault Detection in Digital Circuits*, Prentice-Hall, Englewood Cliffs, N.J., 1971.

[5] Harrison, M. A., *Introduction to Switching and Automata Theory*, McGraw-Hill, New York, N.Y., 1965.

[6] Hayes, J. P., "A NAND Model for Fault Diagnosis in Combinational Logic Networks," *IEEE Trans. on Computers*, vol. C-20, pp. 1496-1506, December, 1971.

[7] Hayes, J. P., "The Fanout Structure of Switching Functions," Switching and Automata Theory Conference, New Orleans, La., 1974.

[8] Klir, G. J., "On Universal Logic Primitives," *IEEE Trans. on Computers*, vol. C-19, pp. 467-469, 1970.

[9] McCluskey, E. J., "Iterative Combinational Switching Networks: General Design Considerations," *IRE Transactions on Electronic Computers*, vol. EC-7, pp. 285-291, December, 1958.

[10] McCluskey, E. J., "Detection of Group Invariance or Total Symmetry of a Boolean Function," *Bell System Technical Journal*, vol. 35, pp. 1445-1453, November, 1956.

[11] McNaughton, R., "Unate Truth Functions," *IRE Trans. Electronic Computers*, vol. EC-10, pp. 1-6, March, 1961.

[12] MacSorley, O. L., "High-Speed Arithmetic in Binary Computers," *Proc. IRE*, vol. 49, pp. 67–91, January, 1961.

[13] Miller, R. E., *Switching Theory: Volume I: Combinational Circuits*, J. Wiley & Sons Inc., New York, N.Y., 1965.

[14] Minnick, R. C., "Magnetic Bubble Computer Systems," *Proc. Fall Joint Computer Conference*, pp. 1279–1298, 1972.

[15] Mukhopadhyay, A. (ed.), *Recent Developments in Switching Theory*, Academic Press, 1970.

[16] Paull, M. C. and E. J. McCluskey, "Boolean Functions Realizable with Single Threshold Devices," *Proc. IRE*, July, 1960.

[17] Post, E. L., "Two-Valued Iterative Systems of Mathematical Logic," *Princeton Univ. Press*, 1941.

## PROBLEMS

**4.1)** For the combinational function $f(x_1, x_2, x_3) = \Sigma\, m_3, m_5, m_6, m_7$ find a factored realization with fan-in index equal to 2.

**4.2)** a) Find a modular realization of a parallel subtractor.
b) Discuss the possible applicability of the techniques for high speed adder design.

**4.3)** Design a modular realization of a circuit which examines a word for the bit patterns consisting of three successive 0's or 1's and generates a 1 output if and only if the word contains such a pattern.

**4.4)** a) Prove that a number $X = x_n x_{n-1}, \ldots, x_1 x_0$ can be incremented by 1 by complementing all bits to the right of and including the least significant 0, assuming $X < 2^{n+1} - 1$.
b) Find a modular realization of this function, and compare its complexity with that of a parallel adder.
c) Find a similar algorithm for decrementing by 1 and a corresponding modular realization?

**4.5)** If a 1-dimensional iterative array corresponds to a sequential circuit to what would a 2-dimensional array (identical cells with 2 outputs interconnected in geometrically uniform rectangular manner) correspond?

**4.6)** Find a minimal two level multiple output realization of the module described by the Karnaugh map of Figure 4.14 and compare it with the cost of the basic cell of a parallel adder.

**4.7)** Prove that any non 0-preserving function is also either non 1-preserving or non self-dual.

**4.8)** For the set of five properties which are required for strong completeness (Theorem 4.3), is it possible for a function to have one of these properties and none of the other four? For each property give an example or a proof.

**4.9)** For each of the following sets of functions determine whether it is
a) strong complete
b) weak complete
c) a strong basis
d) a weak basis
   (i) $\{f_1\}$ where $f_1(x_1,x_2,x_3) = \Sigma\, m_1,m_2,m_4,m_7$
   (ii) $\{f_1,f_2\}$ where $f_2(x_1,x_2,x_3) = \Sigma\, m_0,m_1,m_2,m_7$.
   (iii) $\{f_1,f_2,f_3\}$ where $f_3(x_1,x_2,x_3) = \Sigma\, m_0,m_3,m_4$

**4.10)** Find a single function of 3 variables $f(x_1,x_2,x_3)$, other than the NAND and NOR, which is strong complete.

**4.11)** For what values of $i$ is $S_i(x_1, ..., x_n)$ a threshold function and what are the respective values of the weights and threshold?

**4.12)** Consider the function $T_i(x_1, ..., x_n)$ which has the value 1 if $i$ or more of the variables are 1. For what values of $i$ is $T_i$ a threshold function?

**4.13)** a) Prove that if $x_1$ is adjacent to $x_2$ for $f(x_1,x_2, ...,x_n)$ then $x_1 \sim x_2$ or $x_1 \sim \bar{x}_2$.
b) If $x_1 \sim x_2$ for $f(x_1,x_2, ...,x_n)$ is $x_1$ adjacent to $x_2$?

**4.14)** Prove that $f(\mathbf{x})$ is symmetric if and only if $f(\mathbf{x}) = S_{i_1,i_2,\dots,i_p}(x_1^*, x_2^*, \dots, x_n^*)$.

**†4.15)** Find necessary and sufficient conditions for a set of functions to be
(a) Weak c-complete
(b) Strong c-complete

**4.16)** Prove that a completely specified function $f$ is positive if and only if a minimal sum of products expression for $f$ has no complemented variables.

**4.17)** Prove that the circuits of Figures 4.25(a) and (b) can be made to realize the same function by an appropriate choice of $w_x$ and find the smallest such value of $w_x$ as a function of $T_1, T_2, w_{i1}, w_{i2}$.

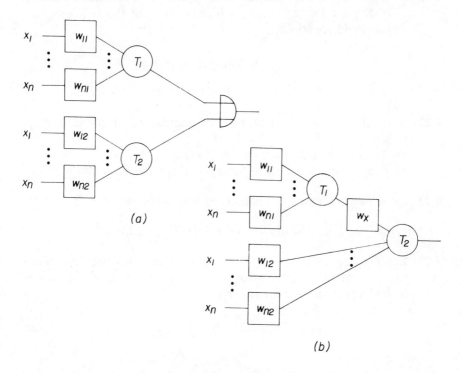

Figure 4.25   Problem 4.17

**4.18)** a) Prove that all prime implicants of a completely specified unate function are essential.

b) Is the above statement true for incompletely specified unate functions? Prove or give a counterexample.

**4.19)** a) Consider a threshold function $f(x_1, x_2, ..., x_n)$ with weights $w_i$, $i = 1, ..., n$, and threshold $T$. If all $w_i$ are equal, $f$ is called a *voting function*. Prove that all voting functions are symmetric.

b) Are all symmetric functions voting functions? If not can you specify constraints on the numbers $k_1, k_2$ ... etc. in order that $S_{k_1, k_2, k_3...}(x_1, x_2, ..., x_n)$ be a voting function.

**4.20)** If $f(x_1, x_2, ..., x_n)$ is a threshold function is the dual of $f$ always a threshold function?

**4.21)** a) If $f(x_1, x_2, ..., x_n)$ is a threshold function and $x_p$ is not a member of the set $\{x_1, x_2, ..., x_n\}$ which, if any, of the following are threshold functions.

$$\bar{x}_p f, \bar{x}_p + f, x_p + f$$

b) Repeat (a) if $x_p$ is a member of the set $\{x_1, x_2, ..., x_n\}$.

**4.22)** A function $f(\mathbf{x})$ is said to be dual comparable if the dual of $f$, $f_d$ is such that $f + f_d = f$ or $f + f_d = f_d$.

a) Are all threshold functions dual comparable?

b) Are all dual comparable functions threshold functions?

**4.23)** How many symmetric functions of $n$ variables are there?

**†4.24)** How many unate functions of $n$ variables are there?

**4.25)** Let $f(x_1, x_2, ..., x_n)$ be a Boolean function. Assume that there exists a set of weights $w_1, w_2, ..., w_n$ ($w_i = +1$ or $-1$, $1 \leq i \leq n$) and a real number $T$ such that

$$f(x_1, x_2, ..., x_n) = 1 \text{ if } \sum_{i=1}^{n} w_i x_i \geq T \qquad \text{and}$$

$$f(x_1, x_2, ..., x_n) = 0 \text{ if } \sum_{i=1}^{n} w_i x_i < T$$

This function is obviously a special type of threshold function. Is $f(x_1,x_2, ...,x_n)$ also a totally symmetric function? If so, prove, and indicate the variables of symmetry and the value of $T$. If not, display some such $f(x_1,x_2, ...,x_n)$ which is not totally symmetric.

**4.26)** The *Fibonacci numbers* are a sequence defined by the relation

$$F_n = F_{n-1} + F_{n-2}$$

where $F_i$ is the $i^{th}$ number in the sequence. The first numbers in the sequence are $1,1,2,3,5,8, \cdots$.
Consider the set of functions

$$F = x_n + x_{n-1}(x_{n-2} + x_{n-3}(x_{n-4} + x_{n-5}(\cdots + x_2(x_1)))) \cdots$$
$$(n \text{ odd})$$

Show that all functions in this set are threshold functions with weights

$$w_n = F_n \text{ (the } n^{th} \text{ Fibonacci number)}$$

and find the value of $T$. Also show that this is true for the dual of $f$ (obtained by interchanging $+$ and $\cdot$) and find the value of $T$.

**4.27)** Which, if any, of the following functions are fanout-free
(a) $f_1 = x_1 x_2 + \bar{x}_1 \bar{x}_3$
(b) $f_2 = x_1 x_2 \bar{x}_3 + x_1 \bar{x}_3 \bar{x}_4$
(c) $f_3 = x_1 \bar{x}_2 \bar{x}_3 x_4 + x_1 \bar{x}_2 x_4 x_5 + x_1 \bar{x}_2 \bar{x}_3 x_5$

# CHAPTER 5

# Sequential Circuits and Machines

## 5.1 SEQUENTIAL FUNCTIONS AND MACHINES

A Boolean function, the value of which depends on previous values of variables as well as the present values, is called a *sequential function*. Whereas a combinational function defines a mapping between inputs and outputs, a sequential function defines a mapping between *input sequences* and *output sequences*. Therefore a sequential function cannot be described by a truth table representation. To describe such functions we will make use of a mathematical model called a *sequential machine*. In this model the effect of all previous inputs on the output is represented by a *state* of the machine. If the set of possible input combinations* is denoted by $I$, the set of possible outputs by $Z$ and the set of possible states by $Q$ the machine operates as follows: If initially in state $q_i \in Q$ the input $I_j \in I$ is applied the output $z_k \in Z$ is generated and the machine goes to state $q_m \in Q$. Both $z_k$ and $q_m$ are uniquely determined from $q_i$ and $I_j$. Thus the output of the machine at any time depends on the state and the input, and these also determine the next state. We shall be interested in sequential functions which can be described by sequential machines with a finite number of states (*finite state machines*). The operation of such a sequential machine can be described by a *state table*. The rows of the state table correspond to the states of the machine, the columns correspond to inputs, and the entry in row $q_i$ and column $I_j$ represents the next state and output for the machine transition caused by applying input $I_j$ to the machine in state $q_i$. The

---

*By input combination we mean a value of a set of binary variables $x_i$.

next state of this transition is denoted by $N(q_i, I_j)$ and the output generated by the transition is denoted by $Z(q_i, I_j)$.

An example of such a state table is shown in Figure 5.1. This machine has four states, $(q_1, q_2, q_3, q_4)$, two possible inputs (0 and 1), and two possible outputs (0 and 1). The state table can be used to determine the output sequence generated for any input sequence and initial state.

$$x$$

|       | 0        | 1        |
|-------|----------|----------|
| $q_1$ | $q_3,0$  | $q_1,0$  |
| $q_2$ | $q_2,0$  | $q_1,0$  |
| $q_3$ | $q_4,1$  | $q_3,1$  |
| $q_4$ | $q_4,0$  | $q_2,0$  |

Figure 5.1   State table of a sequential machine

Thus if the input sequence 0101 is applied to initial state $q_1$ the output sequence 0110 is generated. This is determined in the following manner. The first input in the sequence, which is 0, is applied when the machine is in initial state $q_1$. From the state table entry in row $q_1$ and column 0, we observe that in response to this input a 0 output is generated and the machine goes to state $q_3$. In that state the second input 1 is applied. From row $q_3$ and column 1 we determine that this causes a 1 output and the machine remains in state $q_3$. The 3rd input 0 applied to this state generates a 1 output and the machine goes to state $q_4$. The 4th input, 1, applied to state $q_4$ generates a 0 output and the machine goes to state $q_2$. Thus if the input sequence 0101 is applied in initial state $q_1$, the output sequence 0110 is generated and the final state of the machine is $q_2$. Note that if the same input sequence is applied to initial state $q_4$, a different output sequence, 0000 is generated.

A sequential machine can also be represented in a graph form, called the *state diagram* or *state transition graph*. In a state diagram there is a node corresponding to each state and a directed branch corresponding to each state transition from $q_i$ to $N(q_i, I_j)$ with the input $I_j$ and output $z_k = Z(q_i, I_j)$ indicated by a branch label of the form $I_j / z_k$.

For the machine represented by the state table of Figure 5.1, the state diagram is shown in Figure 5.2.

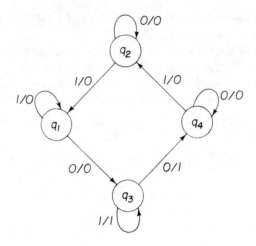

Figure 5.2   State diagram of sequential machine

|       | $I_1$    | $I_2$   | $I_3$   |
|-------|----------|---------|---------|
| $q_1$ | $q_1,0$  | $q_4,0$ | $q_2,0$ |
| $q_2$ | $q_4,1$  | -       | $q_3,0$ |
| $q_3$ | $q_5,1$  | -       | $q_3,1$ |
| $q_4$ | $q_5,1$  | $q_4,1$ | $q_3,1$ |
| $q_5$ | $q_1,0$  | $q_1,0$ | $q_3,1$ |

Figure 5.3   Sequential machine with prohibited input sequences

A state table may also contain (*don't care*) unspecified entries. These represent conditions where the next state and/or the output need not be specified for a given transition from state *i*, input $I_k$, because the input $I_k$ cannot (or is not permitted to) occur when the machine is in state *i*, or because the next state and/or the output are irrelevant or should be ignored when the input $I_k$ occurs in state *i*. For example, in the table of Figure 5.3 the sequential function is undefined for the input sequence $I_3 I_2$ (i.e., this input sequence is prohibited). With initial state $q_1$, $q_4$ or $q_5$ the unspecified transitions in the table occur only for input sequences containing this prohibited subsequence. Thus, these transitions are don't cares.

This model of sequential machines, in which the output depends on both the input and state is called the *Mealy machine model*.[16] In another model of sequential machines, called the *Moore machine model*,[18] the output depends only on the state. In this model if an input $I_j$ is applied to a state $q_i$ the output $Z(q_i)$ which depends only on the initial state $q_i$ is generated and the machine goes to a state $N(q_i, I_j)$ which depends on both $q_i$ and $I_j$. The state table of a Moore machine has an output column with the entry $Z(q_i)$ in row $q_i$. The entry in input column $I_j$ and row $q_i$ is $N(q_i, I_j)$. An example of such a table is shown in Figure 5.4. The input sequence 011 applied to state $q_1$ generates the output sequence $011 = Z(q_1)Z(q_2)Z(q_4)$.

$$x$$

|       | 0     | 1     | $z$ |
|-------|-------|-------|-----|
| $q_1$ | $q_2$ | $q_3$ | 0   |
| $q_2$ | $q_1$ | $q_4$ | 1   |
| $q_3$ | $q_2$ | $q_2$ | 0   |
| $q_4$ | $q_1$ | $q_4$ | 1   |

Figure 5.4  A Moore machine state table

Either of these two sequential machine models can be used to represent a sequential function. For any Moore model machine $M_1$, there exists a Mealy model machine $M_1'$ which is equivalent to $M_1$ in the following sense. For any input sequence $I$ and initial state $q$ of $M_1$, there is a state $q'$ of $M'$ such that $M$ and $M'$ generate the same output sequence in response to $I$, assuming corresponding initial states $q$ and $q'$. Similarly for any Mealy model machine there exists a Moore model machine which is equivalent in the same sense.

For a given Moore model machine $M$ an equivalent Mealy model machine $M'$ can be generated by the following procedure.

**Procedure 5.1**
1) Assume the Moore model machine has the set of possible inputs $I$, the set of possible outputs $Z$, the set of states $Q$ and the next state transition function $N(q_i, I_j)$ and output function $Z(q_i)$. The corresponding Mealy machine has input set $I$, output set $Z$, and state set $Q'$ which has one state $q_i'$ corresponding to each state $q_i$ of $M$.

2) The next state transition function $N'(q_i',I_j)$ of $M'$ is defined as follows: If $N(q_i,I_j) = q_k$ then $N'(q_i',I_j) = q_k'$.

3) The output function $Z'(q_i,I_j)$ of $M'$ is defined as follows: If $N(q_i,I_j) = q_k$ and $Z(q_k) = z_m$ then $Z'(q_i',I_j) = z_m$.

**Example 5.1**

For the Moore machine of Figure 5.4, the equivalent Mealy machine is described by the state table of Figure 5.5. The entry $N(q_2',1)$, $Z(q_2',1)$ is determined from the values of $N(q_2,1)$, $Z(N(q_2,1))$ for the table of Figure 5.3. Since for that table $N(q_2,1) = q_4$ and $Z(q_4) = 1$, $N(q_2',1) = q_4'$ and $Z(q_2',1) = 1$.

$$x$$

|  | 0 | 1 |
|---|---|---|
| $q_1'$ | $q_2',1$ | $q_3',0$ |
| $q_2'$ | $q_1',0$ | $q_4',1$ |
| $q_3'$ | $q_2',1$ | $q_2',1$ |
| $q_4'$ | $q_1',0$ | $q_4',1$ |

**Figure 5.5   An equivalent Mealy machine**

For both the Mealy and Moore models, inputs and outputs can be assumed to occur at quantized instants of time, the initial state occurring at time $t = 0$ and the $i$-th input occurring at time $t = (i - 1)$ and causing a state transition at time $t = i$. Under this interpretation there is one difference between the operation of the Moore and Mealy model machines. This difference is seen by comparing the machines of Figure 5.4 and Figure 5.5. If the input $x = 0$ is applied to the machine of Figure 5.4 in initial state $q_1$ at time $t = 0$, the output $Z(q_1) = 0$ is produced at time $t = 0$, the state changes to $q_2$ at $t = 1$ and the output at $t = 1$ becomes $Z(q_2) = 1$. The corresponding operation of the machine of Figure 5.5 results in the output $Z'(q_1',0) = 1$ at time $t = 0$ and state $q_2'$ is reached at time $t = 1$. Thus the outputs of the Mealy machine occur one time period before the corresponding outputs of the Moore machine, and the initial Moore output must be ignored.* Under this interpretation it can be shown that the Mealy machine derived by Procedure 5.1 is equivalent to the given Moore machine.

*It is possible to derive a Mealy machine whose outputs occur one time period after the corresponding Moore machine outputs. (Problem 5.19)

**Theorem 5.1**

The Mealy machine $M'$ derived by Procedure 5.1 is equivalent (as previously defined) to the given Moore machine $M$.

**Proof**

Suppose $M'$ is not equivalent to $M$. Then there is a state $q_i$ of $M$ and an input sequence $I = I_1 I_2 \ldots I_k$ which generates an output sequence $Z = z_0 z_1 \ldots z_{k-1}$ and no initial state of $M'$ generates the same output sequence for this input sequence. Assume the input sequence $I$ passes $M$ through the state sequence $q_i q_1 q_2 \ldots q_k$ with output sequence $Z$, where $z_j = Z(q_j)$. The same input sequence applied to state $q_i'$ of $M'$ passes $M'$ through the corresponding state sequence with output sequence $z_1' z_2' \ldots z_{k-1}'$ where $z_1' = Z(N(q_i, I_1)) = Z(q_1), z_2' = Z(N(q_1, I_2)) = Z(q_2)$, $\ldots, z_k' = Z(q_k)$. Therefore $M$ in initial state $q_i$ and $M'$ in initial state $q_i'$ generate the same output sequence (ignoring the first output of $M$) in response to input sequence $I$. This contradicts the assumption and proves the theorem.

For a given Mealy machine $M$, the following procedure can be used to define a Moore machine which is equivalent in the sense previously explained.

**Procedure 5.2**

1) Assume the Mealy machine $M$ has the set of possible inputs $I$, the set of possible outputs $Z$, the set of possible states $Q$ and next state function $N(q_i, I_j)$, output function $Z(q_i, I_j)$. The corresponding Moore machine $M'$ has input set $I$, output set $Z$, and state set $Q'$. For each state $q_i$ of $M$ there will be one or more corresponding states $q_{iz_1}', q_{iz_2}' \ldots q_{iz_k}'$, one for each distinct next state/output entry of $M$ of the form $(q_i, z_p)$. If state $q_i$ does not appear as a next state entry in $M$, then $M'$ has one corresponding state $q_i'$.

2) The next state transition function $N'(q_{iz_p}', I_j)$ of $M'$ is defined as follows: If $N(q_i, I_j) = q_k$ and $Z(q_i, I_j) = z_m$ then $N(q_{iz_p}', I_j) = q_{kz_m}'$ (for all $z_p$).

3) The output function $Z'(q_{iz_p}')$ of $M'$ is defined as follows: $Z'(q_{iz_p}') = z_p, Z'(q_j')$ is unspecified.

**Example 5.2**

For the Mealy machine of Figure 5.6(a) the corresponding Moore machine is defined by the state table of Figure 5.6(b). States $q_3$ and

$$x$$
|  | 0 | 1 |
|---|---|---|
| $q_1$ | $q_3,0$ | $q_1,0$ |
| $q_2$ | $q_2,0$ | $q_1,0$ |
| $q_3$ | $q_4,1$ | $q_3,1$ |
| $q_4$ | $q_4,0$ | $q_2,0$ |
| $q_5$ | $q_3,1$ | $q_1,0$ |

(a)

$$x$$
|  | 0 | 1 | $z$ |
|---|---|---|---|
| $q_{10}$ | $q_{30}$ | $q_{10}$ | 0 |
| $q_{20}$ | $q_{20}$ | $q_{10}$ | 0 |
| $q_{30}$ | $q_{41}$ | $q_{31}$ | 0 |
| $q_{31}$ | $q_{41}$ | $q_{31}$ | 1 |
| $q_{40}$ | $q_{40}$ | $q_{20}$ | 0 |
| $q_{41}$ | $q_{40}$ | $q_{20}$ | 1 |
| $q_5'$ | $q_{31}$ | $q_{10}$ | - |

(b)

Figure 5.6   (a) A Mealy machine, (b) An equivalent Moore machine

$q_4$ appear as next state entries with outputs $Z = 0$ and $Z = 1$ so there are two corresponding states $(q_{30},q_{31})$, $(q_{40},q_{41})$, for each. State $q_5$ does not appear as a next state entry in $M$. The corresponding state of $M'$ is $q_5'$. The output $Z(q_{ij}) = j$ for $j = 0,1$. The next state transition $N'(q_{41},1) = q_{iz_1}$ where $N(q_4,1) = q_i$ and $Z(q_4,1) = z_1$. Thus $N'(q_{41},1) = q_{20}$.

## Theorem 5.2

The Moore machine $M'$ derived by Procedure 5.2 is equivalent (as previously defined) to the given Mealy machine $M$.

**Proof:**
Similar to proof of Theorem 5.1. Exercise.

## 5.2 SEQUENTIAL CIRCUITS

### 5.2.1 MEMORY ELEMENTS

Physical realizations of sequential machines are called *sequential circuits*. To represent the state of a sequential machine, memory elements are required. We shall now consider several different types of memory elements. These devices generally have two stable conditions which can be represented as 0 or 1 states. Such memory elements are called *bistable multivibrators* or *flip-flops (FF)*.

An *SR (Set-Reset) FF* can be realized by the circuit of Figure 5.7(a). It has two inputs $S$ and $R$, to gates $G_1$ and $G_2$ respectively. The outputs of these gates shall be denoted as $y_1$ and $y_2$ respectively. We wish this circuit to operate in such a manner that the two stable configurations are $y_1 = 1, y_2 = 0$ and $y_1 = 0, y_2 = 1$. For the input $S = R = 1$ the outputs $y_2$ and $y_1$ are both 0. Hence this input is always prohibited.

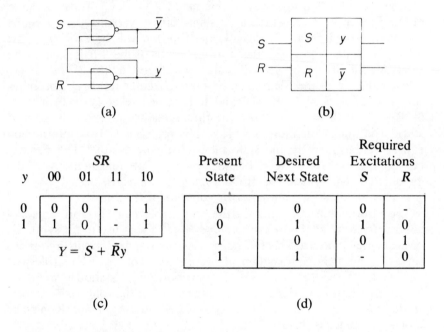

(a)                                    (b)

| $y$ | $00$ | $01$ | $11$ | $10$ |
|-----|------|------|------|------|
| 0   | 0    | 0    | -    | 1    |
| 1   | 1    | 0    | -    | 1    |

$SR$

$$Y = S + \bar{R}y$$

| Present State | Desired Next State | Required Excitations $S$ | $R$ |
|---------------|--------------------|--------------------------|-----|
| 0 | 0 | 0 | - |
| 0 | 1 | 1 | 0 |
| 1 | 0 | 0 | 1 |
| 1 | 1 | - | 0 |

(c)                                    (d)

**Figure 5.7** (a) **Realization, (b) representation, (c) state table, (d) transition table for an** *SR* **flip-flop**

If $S = 1$ and $R = 0$, the output of $G_1, y_1$, becomes 0 independent of the value of $y_2$. After $y_1$ becomes 0 the output of gate $G_2$ becomes 1 (since $y_1$ is an input to $G_2$). The memory element remains in this configuration $y_1 = 0, y_2 = 1$, until the inputs change. If the inputs are $S = 0$ and $R = 1$ the output of gate $G_2$ becomes 0 which causes the output of gate $G_1$ to become 1. This configuration $y_1 = 1, y_2 = 0$, is also stable. For the inputs $S = 1$, $R = 0$ or $S = 0$, $R = 1$ the stable configuration reached does not depend on the state of the memory element when the inputs are applied. However if $S = R = 0$ the stable configuration reached does depend on this initial state. Suppose the FF is stable with $y_1 = 1, y_2 = 0$ when the inputs $S = R = 0$ are applied. Since $y_1 = 1$ the output of gate $G_2$ remains 0. Since both inputs to $G_1$ are 0, the output $y_1$ remains 1. Hence the FF remains in the same stable state $y_1 = 1, y_2 = 0$. Similarly for the initial state $y_1 = 0, y_2 = 1$, and $S = R = 0$, the final stable state would remain $y_1 = 0, y_2 = 1$.

If the SR flip-flop operates in this manner $y_2 = \bar{y}_1$. Thus the state of the FF can be represented by a single binary variable $y_2 = y$. Since the input $S = 1$ results in $y = 1$, the flip-flop will be said to be *set* if it is in state $y = 1$. Similarly in state $y = 0$ the FF is said to be *reset*. An SR FF is represented as shown in Figure 5.7(b).

The behavior of the SR FF can be summarized as follows: it becomes set $(y = 1)$ if $S = 1$, $R = 0$ and it becomes reset $(y = 0)$ if $S = 0$, $R = 1$. If $S = R = 0$ the flip-flop remains stable in its present state. The input condition $S = R = 1$ is prohibited. This information can be represented by the state table of Figure 5.7(c).* The FF will be set if $S = 1$ or if $R = 0$ and $y = 1$. If we represent by $Y$ the condition under which the FF becomes (or remains) set, $Y = S + \bar{R}y$. This is called the *characteristic equation* of the FF.

The information contained in the state table can be represented in another form called the *transition table* which is very useful for synthesis of sequential circuits. This table specifies the required input *excitations* for every possible combination of present state of the FF and desired next state of the FF. The SR-FF transition table is shown in Figure 5.7(d). The first row of this table is interpreted in the following manner. If the FF is to remain stable at $y = 0$, $S$ must be 0 but $R$ may be 0 or 1 and hence is unspecified.

A second type of memory element is the *JK flip flop*. This operates in a similar manner to the SR flip flop, the input $J$ being used to set the element and the input $K$ to reset it. However the input condition $J = K = 1$ is permitted and results in the device changing state, i.e., if the present state is $y = 1$ the state becomes $y = 0$, and vice versa. The realization, representation and behavior of this element is shown in Figure 5.8. If the input $J = K = 1$ persists for a sufficiently long period the FF may oscillate. A clock enabling signal (to be discussed in Section 5.2.2) is used to prevent this.

---

*In the state tables for FF's, we will represent the states simply as 0 and 1 rather than as $q_0$ and $q_1$.

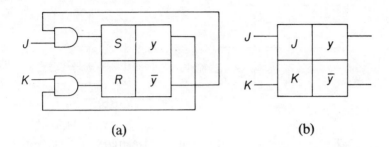

| | JK | | | | Present | Desired | Required Excitations | |
|---|---|---|---|---|---|---|---|---|
| y | 00 | 01 | 11 | 10 | State | Next State | J | K |
| 0 | 0 | 0 | 1 | 1 | 0 | 0 | 0 | - |
| 1 | 1 | 0 | 0 | 1 | 0 | 1 | 1 | - |
| | | | | | 1 | 0 | - | 1 |
| | $Y = J\bar{y} + \bar{K}y$ | | | | 1 | 1 | - | 0 |

(c)                                              (d)

**Figure 5.8   (a) Realization, (b) representation, (c) state table, (d) transition table of *JK* FF**

The *T (trigger) flip-flop* (Figure 5.9) operates like a *JK* flip-flop if the inputs *J* and *K* are constrained to always be equal. In this case these two inputs can be replaced by a single input *T*. If $T = 0$ the device remains in its present state, and if $T = 1$ the device changes state.

The *D (delay) FF* (which is also called a *delay element*) behaves as an *SR* flip-flop if the inputs *S* and *R* are constrained to be complementary (i.e. $S = \bar{R}$), and replaced by a single input *D*. If $D = 1$ the device becomes set and if $D = 0$ the device becomes reset. Hence the state of a *D FF* is always equal to the value of the previous excitation. The behavior of this element is as represented in Figure 5.10.

In practice gates also have delay associated with them. That is there is a non-zero elapsed time between an input change to a gate and the output change produced by the input change. The effect of this delay can be modeled by a *delay element of magnitude D*. If the input to

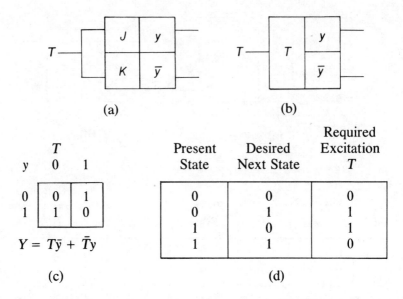

Figure 5.9   (a) Realization, (b) representation, (c) state table, (d) transition table of *T FF*

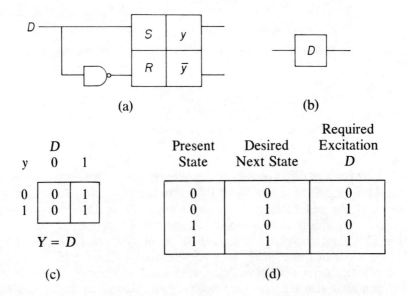

Figure 5.10   (a) Realization, (b) representation, (c) state table, (d) transition table of *D FF*

the delay element is described by the Boolean time function $x(t)$, and the output by $y(t)$, then $y(t) = x(t - D)$ is the characteristic equation of a delay element of magnitude $D$. A $D$-$FF$ operates in a similar manner to a delay element.

## 5.2.2 DELAYS AND CLOCKS

In practice there is a delay between signal changes on the inputs of a gate and the consequent changes on the gate output. Until now we have ignored the effect of gate delays on the behavior of circuits constructed from gates, since this effect is inconsequential for combinational circuits. However the consequences for sequential circuits can be serious. Consider the combinational circuit of Figure 5.11(a). If gates $G_1, G_2, G_3$ have associated delays of magnitudes $\Delta_1, \Delta_2, \Delta_3$ respectively, this circuit can be modelled as shown in Figure 5.11(b) where the gates $G'_1, G'_2, G'_3$ are assumed to be idealized zero delay gates. Suppose that at time $t_o$ the inputs are $x_1 = x_2 = x_3 = 1$, and at time $t > t_o$, $x_1$ changes to 0 while $x_2$ and $x_3$ remain at 1. Since $f(1,1,1) = f(0,1,1) = 0$ we would expect the output of the circuit to remain stable at 0. However the output of gate $G_1$ will change from 1 to 0 and the output of gate

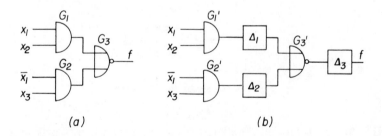

Figure 5.11   (a) A combinational circuit and (b) its gate delay model

$G_2$ will change from 0 to 1. If $\Delta_1 < \Delta_2$ then the $G_1$ change will occur before the $G_2$ change. In the interval between the $G_1$ change and the $G_2$ change both inputs to $G_3$ are 0. Figure 5.12 illustrates the signals $x_1, G_1, G_2, G_3$ as functions of time. From this figure it is apparent that if $\Delta_1 < \Delta_2$, even by a very small amount $\epsilon$, the circuit output will contain a 1-pulse of duration $\epsilon$.

We thus see that due to delays a combinational circuit may produce a transient error. If the output of this circuit is an input to a flip-flop y, the erroneous transient 1-pulse may result in incorrectly setting or resetting y. The effect of this is a permanent or stable error rather than a transient error. In general a flip-flop will not respond to pulses of duration less than some minimum value called the *trigger time*. Hence the pulse width must exceed the trigger time to produce such an error.

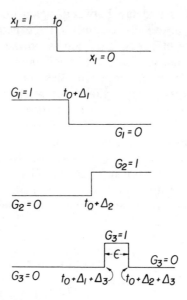

**Figure 5.12    Signal timing diagram**

This problem can be eliminated by using a clock, which is assumed to be generated independently. The clock signal $C$ can be used so that a flip-flop will only change state when $C = 1$ as shown in Figure 5.13. In this way transient *FF* excitations can be *masked* so as not to affect the circuit operation. Sequential circuits which utilize clocks in this manner are called *synchronous* circuits. Sequential circuits can also be designed *asynchronously* without clocks[5,21]. However we shall only consider synchronous circuits.

If the clock signal $C$ has a period $T$, during which the clock pulse duration (width) (i.e. the time when $C = 1$) is $T_p$ and $C = 0$ for time $T - T_p$, then $T$ and $T_p$ must be constrained in terms of the gate delays and the memory element trigger times to ensure proper circuit operation. Specifically $T - T_p$ must be sufficiently large to permit all $FF$ excitations to stabilize, (i.e. $C = 0$ during the period that $FF$ excitation inputs may be changing value). Similarly $T_p$ must be larger than the trigger time of any $FF$. The clock pulse width $T_p$ also has an upper bound. Specifically if as in Figure 5.14, the outputs of a $FF$ $y$ are inputs to

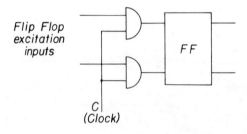

Figure 5.13   Clocked flip-flop

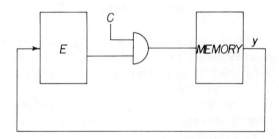

Figure 5.14   Circuit with clocked flip-flop excitation

a circuit $E$ which generates the excitation inputs to some other $FF$ $y'$ (or to $y$ itself) the clock pulse width $T_p$ must be such that $FF$ $y'$ (or $y$) does not respond to the change in $y$ until the next clock period. Thus if the delay of the circuit $E$ is $\Delta E$, the delay of $FF$ $y$ is $\Delta y$ and the minimum trigger time of a *FF* is $\Delta s$ then $T_p < \Delta E + \Delta y + \Delta s.*$ Thus the maximum clock pulse width must be constrained.

Alternatively proper operation can be ensured by using a *master-slave flip-flop* (Figure 5.15), which consists of two flip flops connected in cascade. The first flip flop can only change state if $C = 1$ while the second flip flop changes to the same state as the first flip flop when $C = 0$. If a master-slave *FF* is utilized in a circuit such as that of Figure 5.14, the inputs to circuit $E$ are taken from the outputs of the second *FF*. Therefore these inputs cannot change when $C = 1$. When

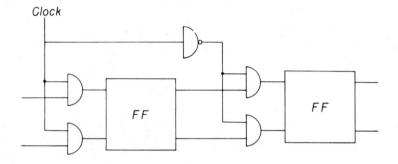

Figure 5.15   Master-slave *FF*

$C = 0$ these inputs may change which may then cause the outputs of $E$ to change. However these changes cannot affect the state of the master-slave *FF* until $C = 1$. Thus the maximum width of the clock pulse need not be restricted.

In designing synchronous circuits, we shall assume that the clock has been properly designed or master-slave *FF*'s have been utilized to ensure proper operation.


## 5.3 SYNTHESIS OF SEQUENTIAL CIRCUITS

The synthesis of a sequential circuit consists of designing a sequential circuit which performs the computation specified in some general de-

---

*Note that for a *JK-FF* if $J = K = 1$, the state of the *FF* will always oscillate if the clock pulse is sufficiently long.

scription of a sequential function. This process consists of several distinct problems which can be represented by the block diagram of Figure 5.16.

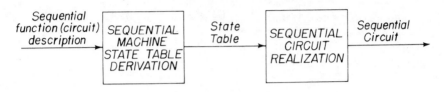

Figure 5.16   Sequential circuit synthesis

## 5.3.1 STATE TABLE DERIVATION

There is no algorithmic procedure for the state table derivation step in the synthesis of sequential circuits. The major difficulty is in deciding what information must be remembered and how this information is to be represented as states of a sequential machine. Some types of computation can be formulated as either basically combinational or basically sequential processes. An example of such a computation is binary addition. In the previous chapter we showed how combinational circuit adders could be designed. Addition can also be formulated as a sequential process in the following manner. We assume that there are two binary inputs $x$ and $y$. The sequence of $x$-inputs is interpreted as the binary representation of a single number $X$, low order bits first. Similarly the sequence of $y$-inputs is interpreted as the binary representation of a number $Y$. The binary output $z$, should generate a sequence equal to the binary representation of $X + Y$, low order bits being generated first. (The addition is effectively terminated by having the most significant bits of $X$ and $Y$ equal to 0.)

The essential information which must be remembered when performing the addition of the $i$-th bits of $X$ and $Y$ is the carry from the addition of bit $(i - 1)$ of $X$ and $Y$. Thus this sequential addition process can be represented by a sequential machine with two states, $q_0$ indicating a 0-carry and $q_1$ indicating a 1-carry. Since the initial carry is 0, the initial state of the machine is $q_0$. The state table of this sequential machine which is sometimes referred to as a *serial adder*, is shown in Figure 5.17. The entries of this table are determined in the following manner. A carry is generated (next state entry $q_1$) if the inputs $x$ and

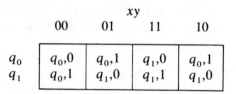

$$
\begin{array}{c}
 & \multicolumn{4}{c}{xy} \\
 & 00 & 01 & 11 & 10 \\
\end{array}
$$

| | 00 | 01 | 11 | 10 |
|---|---|---|---|---|
| $q_0$ | $q_0,0$ | $q_0,1$ | $q_1,0$ | $q_0,1$ |
| $q_1$ | $q_0,1$ | $q_1,0$ | $q_1,1$ | $q_1,0$ |

**Figure 5.17   Serial Adder**

$y$ are both 1 or if there is a carry from the previous bit addition (present state $q_1$) and $x = 0$, $y = 1$ or $x = 1$, $y = 0$. This determines all of the next state entries. The output is 1 if there is a carry from the previous bit addition (present state $q_1$) and the inputs are $x = y = 0$ or $x = y = 1$ or if there is no carry (present state $q_1$) and $x = 1$, $y = 0$ or $x = 0$, $y = 1$.

We shall see that this sequential machine can be realized by a sequential circuit which is very similar to a single module of the combinational parallel adder of Figure 4.12(b). In general, performing a computation sequentially results in a time sharing of hardware and thus a more economical, but slower, circuit than the same computation performed combinationally.

Sequential circuits are also used for computations which cannot be realized by combinational circuits. The following examples demonstrate the state table generation process for such sequential functions.

**Example 5.3**

We wish to design a sequential circuit with one input $x$, and one output $z$, such that $z = 1$ at the $i$-th clock pulse if and only if the $i$-th input is 1 and the previous two inputs were 0. The essential information which must be remembered by the circuit is the information about the past two inputs, specifically were the last two inputs 0, was only the last one input 0 or was the last input 1. To remember this information requires three states.

$q_0$ which represents the fact that the last input was not 0 so that the current string of consecutive 0's is of length 0

$q_1$ which represents the fact that the current string of 0's is of length 1

$q_2$ current string of 0's is two (or more).

We are now at a stage where the state table can be derived. The table will have three rows (corresponding to states $q_0, q_1, q_2$) and two columns (corresponding to $x = 0$, $x = 1$). The entries in the table are specified as follows. The output is 1 only if the present input is 1 and the past two inputs were 0 (which is represented by state $q_2$). Thus, the output entries are as shown in Figure 5.18(a). If the present input is 1, at the next instant of time we will have a string of no consecutive 0's and therefore the next state should be $q_0$ (regardless of the present state, Figure 5.18(b)). If the present input is 0 and the present state indicates $i$ consecutive 0's, the next state should indicate $i + 1$ consecutive 0's for $i < 2$. If $i = 2$ the next state should indicate two consecutive 0's. This enables the state table to be completed, Figure 5.18(c).

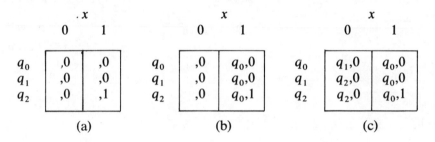

Figure 5.18   State table generation

## Example 5.4

At the intersection of Avenue $A$ and 1st Street there is a traffic light which is to be operated by a traffic controller ($TC$). The controller has a clock $C$ with a one-minute period. The lights should operate so as to alternately allow traffic to flow on Avenue $A$ for one minute and then on 1st Street for one minute. If a pedestrian presses a button on the corner all lights should turn red at the end of the current one minute period, stay red for two minutes, and then resume operation as if the interruption had not occurred, i.e., in the proper sequence.

We wish to realize the controller as a synchronous sequential circuit. This requires that the inputs to the circuit change only when the clock is 0. The only input to $TC$ is the information whether the pedestrian button has been pushed. Since the pedestrian button cannot be assumed to be pushed only when the clock is 0, we shall use the basic system shown in Figure 5.19. We assume that the pedestrian button sets an unclocked $JK$ flip-flop, which is "gated into" an $SR$ flip-flop when the clock is 0. The latter flip-flop provides the input $p$ to $TC$. We can now formulate $TC$ as a (synchronous) sequential circuit.

The circuit realizing $TC$ must produce an output $z_1 = 1$ for traffic flow on Avenue $A$, $z_2 = 1$ for traffic flow on 1st Street. We shall assume that $z_1$ and $z_2$ are inputs to flip-flops which control the traffic lights. A third output $z_3$ is used for resetting the $JK$ flip-flop after servicing it.

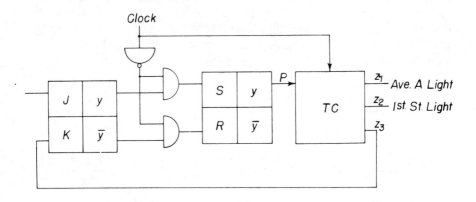

**Figure 5.19    Traffic controller circuit**

The traffic controller can be defined by a sequential machine with four states as follows.

| State | Description |
|---|---|
| $q_1$ | Output $(z_1, z_2, z_3) = 100$, next output = 010, if no interrupt. |
| $q_2$ | Output = 010; next output = 100, if no interrupt. |
| $q_3$ | Reached if interrupt occurs in state $q_1$. |
| $q_4$ | Reached if interrupt occurs in state $q_2$. |

We assume that the initial state is $q_1$. During normal operation (i.e. $p = 0$) the machines alternate between states $q_1$ and $q_2$ generating the output (100) and (010), respectively, as shown in Figure 5.20(a). If an interrupt ($p = 1$) occurs in state $q_1$ the next two outputs should have $z_1 = 0$, $z_2 = 0$. Upon resumption of normal operation, the next output generated should have $z_1 = 1$, $z_2 = 0$. In order to return to normal operation after two clock periods the $JK$ flip flop must be reset after one clock period in the interrupt cycle. This is achieved by generating

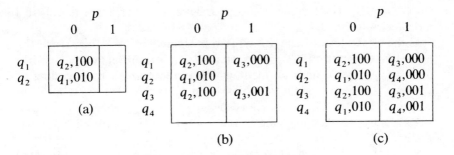

Figure 5.20    Traffic controller state table generation

an output $z_3 = 1$. This behavior is obtained by the state table of Figure
5.20(b). Note that a pedestrian request for service which occurs exactly
when $z_3 = 1$ will be ignored and will not prevent the $JK$ flip-flop from
being reset. An interrupt which occurs in state $q_2$ is handled similarly
and results in resumed normal operation outputs $z_1 = 0$, $z_2 = 1$ as shown
in Figure 5.20(c). Note that the traffic controller has partial control of
its input in that a transition from $p = 1$ to $p = 0$ is initiated and controlled
by the $TC$ circuit. However the transition from $p = 0$ to $p = 1$ is
controlled by an external input.

## 5.3.2 SEQUENTIAL CIRCUIT REALIZATION

The problem of deriving a sequential circuit to realize a sequential
machine specified by a state table is a well defined and relatively simple
process. We assume that the type of memory elements to be used in
the circuit has been specified. The first step in the sequential circuit
derivation is called the *state (variable) assignment problem.* Since the
memory elements are two-state devices, a binary variable $y_i$, called
a *state variable,* is used to denote the state of each memory element.
For a set of state variables if each variable is specified to be 0 or
1, a *state (variable) coding* is obtained. For a set of $k$ state variables
$y = (y_1, y_2, ..., y_k)$ there are $2^k$ different codings. The assignment of
these codings to the states of a sequential machine is called a *state
assignment.* A state assignment must assign at least one coding to each
state and no coding can be assigned to more than one state. Thus an
$n$-state machine requires at least $\lceil \log_2 n \rceil$ memory elements, where $\lceil x \rceil$
is the smallest integer $\geqq x$, in order to obtain $n$ distinct codings. A state

assignment in which one coding is assigned to each state is called a *unicode assignment*.

If the next state of the *FF* $y_i$ is represented by $Y_i$, then since the next state depends on the present state which is represented by a coding of $\mathbf{y} = (y_1, ..., y_k)$, and the input which is represented by a coding of $\mathbf{x} = (x_1, ..., x_m)$ (the input variables) then $Y_i = f_i(\mathbf{x}, \mathbf{y})$, $1 \leqq i \leqq k$. Similarly if the output variable set is $\mathbf{Z} = (z_1, z_2, ..., z_q)$ then $z_i = g_i(\mathbf{x}, \mathbf{y})$, $1 \leqq i \leqq q$ for a Mealy machine and $z_i = g_i(\mathbf{y})$, $1 \leqq i \leqq q$ for a Moore machine.

A synchronous sequential circuit can be schematically represented as shown in Figure 5.21. The basic elements in a synchronous circuit are the clock to ensure proper operation, the memory elements to represent the state, and the combinational circuit to generate the *FF* excitation inputs $Y_E$ and the circuit outputs $z_i$. Once the state assignment has been specified, the derivation of this combinational logic will complete

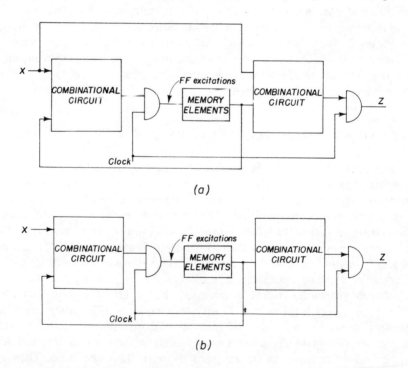

Figure 5.21   Schematic representation of synchronous sequential circuit (a) Mealy model, (b) Moore model

the sequential circuit design. To generate the $FF$ excitation logic it is necessary to determine the value of $Y_E$ for every state transition. This problem can be divided into two parts, the determination of the next state vector $Y$ for each transition, and the determination of the excitations $Y_E$ from $Y$. The next values of the state variables, denoted by $Y_i$, can be obtained directly from the state table. It is convenient to represent $Y_i$ on a Karnaugh map in the variables $\mathbf{y} = (y_1, ..., y_k)$ and $\mathbf{x} = (x_1, ..., x_n)$ (this simplifies the subsequent problem of deriving the logic for the combinational circuit). This map, which specifies the set of functions $Y = f_i(\mathbf{x}, \mathbf{y})$, is called a $Y$-matrix. The transfer of information from the state table to the $Y$-matrix is simplified if each row of the $Y$-matrix represents a state variable coding and each column represents an input variable coding. The entry in a row of the $Y$-matrix corresponding to a coding for state $q_i$ and input $I_k$ of a sequential machine $M$, must correspond to the coding for $N(q_i, I_k)$. If $2^k > n$, some codings may not be assigned to any state. The entries in the $Y$-matrix for the corresponding rows are left unspecified. Since the outputs are also functions of the inputs $\mathbf{x}$ and the state variables $\mathbf{y}$, it is possible to use the same Karnaugh map to represent the functions $z_i = g_i(\mathbf{x}, \mathbf{y})$.* The resulting map is called a $Y$-$Z$ matrix.

## Example 5.5

For the state table and state assignment of Figure 5.22, the $Y$-$Z$ matrix can be represented in the form of a Karnaugh map (Figure 5.23(a))

| | $x$ | | | State Assignment | | |
|---|---|---|---|---|---|---|
| | 0 | 1 | | $y_1$ | $y_2$ | $y_3$ |
| $q_1$ | $q_2,0$ | $q_3,1$ | | 1 | 0 | 0 | 0 |
| $q_2$ | $q_1,0$ | $q_2,0$ | | 2 | 0 | 0 | 1 |
| $q_3$ | $q_5,1$ | $q_3,1$ | | 3 | 1 | 0 | 1 |
| $q_4$ | $q_4,1$ | $q_5,1$ | | 4 | 1 | 0 | 0 |
| $q_5$ | $q_4,0$ | $q_1,0$ | | 5 | 1 | 1 | 1 |
| | (a) | | | | (b) | |

Figure 5.22    State table and state variable assignment

*For a Moore machine the outputs can be specified on a map in the state variables $y_i$.

$x$

| | $y_1$ $y_2$ $y_3$ | 0 | 1 |
|---|---|---|---|
| $(q_1)$ | 0 0 0 | | |
| $(q_2)$ | 0 0 1 | | |
| | 0 1 1 | | |
| | 0 1 0 | | |
| | 1 1 0 | | |
| $(q_5)$ | 1 1 1 | | |
| $(q_3)$ | 1 0 1 | | |
| $(q_4)$ | 1 0 0 | | |

(a)

$x$

| | $y_1$ $y_2$ $y_3$ | 0 | 1 |
|---|---|---|---|
| $(q_1)$ | 0 0 0 | 0010 | 1011 |
| $(q_2)$ | 0 0 1 | 0000 | 0010 |
| | 0 1 1 | ---- | ---- |
| | 0 1 0 | ---- | ---- |
| | 1 1 0 | ---- | ---- |
| $(q_5)$ | 1 1 1 | 1000 | 0000 |
| $(q_3)$ | 1 0 1 | 1111 | 1011 |
| $(q_4)$ | 1 0 0 | 1001 | 1111 |

$Y_1, Y_2, Y_3, Z$

(b)

Figure 5.23   $Y$-$Z$ matrix

with 8 rows, one for each of the state codings, and 2 columns for the inputs $x = 0$, $x = 1$. The state to which each coding is assigned is indicated in parentheses on the left side of the Karnaugh map. The entries of the matrix can be determined from the corresponding state table transitions and state assignment. For instance the entry in row 100 (which is assigned to state $q_4$) and column 1 is the coding for the state $N(q_4, 1) = q_5$ which is 111, and the associated transition output $z = 1$. For those codings which are not assigned to any state (010, 011, 110) the entries are unspecified. The completed $Y$-$Z$ matrix is shown in Figure 5.23(b).

The $Y$-(and $Y$-$Z$) matrix specifies the next value for each state variable $Y_i$ for each transition. The required FF excitations for the corresponding transition (from $y_i$ to $Y_i$) can be determined from the $Y$ matrix and the transition table of the particular memory element type as specified in Figures 5.7–5.10. The Karnaugh map specifying these excitations as functions of the variable sets $\mathbf{y}$ and $\mathbf{x}$ is called the *excitation matrix*.

For the different types of flip-flops described in the previous section the excitation inputs are determined from the value of $Y_i$ and $y_i$ for a given transition. For delay (FF's) elements the excitation matrix is identical to the $Y$ matrix, since $Y_i = D_i$. For $T$ flip flops, $T_i = 1$ for all transitions in which $y_i$ changes value (i.e. the next value, $Y_i$, is not equal to the present value $y_i$) and $T_i = 0$ for all transitions in which $Y_i = y_i$. For $SR$ flip flops $S_i = 1$ and $R_i = 0$ for all transitions in which $y_i = 0$ and $Y_i = 1$, $S_i = 0$ and $R_i = 1$ for all transitions in which $y_i = 1$ and $Y_i = 0$. If $y_i = Y_i = 0$, then $S_i = 0$ and $R_i$ is unspecified, and if $y_i = Y_i = 1$, $S_i$ is unspecified and $R_i = 0$. For $JK$ flip flops if $y_i = 0$ and $Y_i = 1$, $J_i = 1$ and $K_i$ is unspecified, if $y_i = Y_i = 0$, $J_i = 0$ and $K_i$ is unspecified, if $y_i = 1$ and $Y_i = 0$, then $J_i$ is unspecified and $K_i = 1$, and if $y_i = Y_i = 1$, then $J_i$ is unspecified and $K_i = 0$.

**Example 5.6**

The $Y$-matrix of Figure 5.23(b) is transformed into the excitation matrix of Figure 5.24 assuming $T$ flip flops.* The entry in row 111 and column 0 of the $Y$-matrix has next state coding 100. For this transition, the value of state variable $y_1$ stays fixed at 1 ($y_1 = Y_1 = 1$), therefore $T_1$ must be 0. Similarly $y_2 = 1$ and $Y_2 = 0$ implies $T_2 = 1$, and $y_3 = 1$ and $Y_3 = 0$ implies $T_3 = 1$. The corresponding entry (row 111, column 0) of the map of Figure 5.24 specifies these excitations. The other entries are derived in a similar manner. Using the procedures of Section 3.2,

|         |       |       | $x$   |       |
|---------|-------|-------|-------|-------|
| $y_1$ | $y_2$ | $y_3$ | 0     | 1     |
| 0 | 0 | 0 | 001 | 101 |
| 0 | 0 | 1 | 001 | 000 |
| 0 | 1 | 1 | --- | --- |
| 0 | 1 | 0 | --- | --- |
| 1 | 1 | 0 | --- | --- |
| 1 | 1 | 1 | 011 | 111 |
| 1 | 0 | 1 | 010 | 000 |
| 1 | 0 | 0 | 000 | 011 |

$$T_1, T_2, T_3$$

**Figure 5.24    *T FF* excitation matrix**

*In this matrix all three excitation functions have been specified on a single Karnaugh map. Since this map will be used to derive minimal two level realizations of the functions, it may be more convenient to represent each function on a separate Karnaugh map.

we can derive minimal sum of products expressions for each of the excitations considered individually, resulting in

$$T_1 = x\bar{y}_1\,\bar{y}_3 + xy_2$$
$$T_2 = y_2 + \bar{x}y_1\,y_3 + xy_1\,\bar{y}_3$$
$$T_3 = y_2 + x\bar{y}_3 + \bar{x}\bar{y}_1.$$

If sharing is permitted $T_1, T_2$, and $T_3$ can be considered as a multi-output function and the procedures of Section 3.3 can be applied to derive minimal cost solutions for the set of excitation functions.

The output function $z$ does not depend on the type of memory elements being used. This function can be determined from the $Y$-$Z$-matrix of Figure 5.23(b) as

$$z = y_1\,\bar{y}_2 + x\bar{y}_3.$$

For a Moore machine the output functions can be determined from

| $y_1$ $y_2$ $y_3$ | $x$ 0 | 1 | | $y_1$ $y_2$ $y_3$ | $x$ 0 | 1 |
|---|---|---|---|---|---|---|
| 0 0 0 | 001 | 101 | | 0 0 0 | - -0 | 0-0 |
| 0 0 1 | 000 | 00- | | 0 0 1 | - -1 | - -0 |
| 0 1 1 | - - - | - - - | | 0 1 1 | - - - | - - - |
| 0 1 0 | - - - | - - - | | 0 1 0 | - - - | - - - |
| 1 1 0 | - - - | - - - | | 1 1 0 | - - - | - - - |
| 1 1 1 | -00 | 000 | | 1 1 1 | 011 | 111 |
| 1 0 1 | -1- | -0- | | 1 0 1 | 000 | 0-0 |
| 1 0 0 | -00 | -11 | | 1 0 0 | 0- - | 000 |

$$S_1, S_2, S_3$$

$$S_1 = x\bar{y}_3$$
$$S_2 = xy_1\,\bar{y}_3 + \bar{x}y_1\,\bar{y}_2y_3$$
$$S_3 = \bar{y}_1\bar{y}_3 + x\bar{y}_2$$

(a)

$$R_1, R_2, R_3$$

$$R_1 = xy_2$$
$$R_2 = y_2$$
$$R_3 = y_2 + \bar{x}\bar{y}_1y_3$$

(b)

Figure 5.25 SR FF excitation matrix

Karnaugh maps in the state variables $y_i$.

The excitation matrices for $SR$ flip-flops and $JK$-flip flops corresponding to the $Y$-matrix of Figure 5.23(b) are shown in Figures 5.25 and 5.26 respectively with minimal sum of products expressions for each of the excitations.

If delay elements or $D$-$FF$'s are used the excitation matrix is identical to the $Y$-matrix of Figure 5.23(b) and the excitation equations are:

$$D_1 = \bar{x}y_1 + y_1\bar{y}_2 + x\bar{y}_3$$
$$D_2 = \bar{x}y_1\bar{y}_2y_3 + xy_1\bar{y}_3$$
$$D_3 = \bar{y}_1\bar{y}_2\bar{y}_3 + x\bar{y}_2 + y_1\bar{y}_2y_3.$$

For the sequential adder, if $D$-$FF$'s are used as memory elements, the excitation logic circuit is identical to the logic of a single module in the combinational parallel adder. This relation between one dimensional iterative arrays and corresponding sequential circuits is always valid.

With some experience it becomes relatively easy to derive the excitation

| $y_1$ | $y_2$ | $y_3$ | $x$ 0 | 1 |
|---|---|---|---|---|
| 0 | 0 | 0 | 001 | 101 |
| 0 | 0 | 1 | 00- | 00- |
| 0 | 1 | 1 | --- | --- |
| 0 | 1 | 0 | --- | --- |
| 1 | 1 | 0 | --- | --- |
| 1 | 1 | 1 | --- | --- |
| 1 | 0 | 1 | -1- | -0- |
| 1 | 0 | 0 | -00 | -11 |

$$J_1, J_2, J_3$$

$$J_1 = x\bar{y}_3$$
$$J_2 = xy_1\bar{y}_3 + \bar{x}y_1y_3$$
$$J_3 = x + \bar{y}_1$$

| $y_1$ | $y_2$ | $y_3$ | $x$ 0 | 1 |
|---|---|---|---|---|
| 0 | 0 | 0 | --- | --- |
| 0 | 0 | 1 | --1 | --0 |
| 0 | 1 | 1 | --- | --- |
| 0 | 1 | 0 | --- | --- |
| 1 | 1 | 0 | --- | --- |
| 1 | 1 | 1 | 011 | 111 |
| 1 | 0 | 1 | 0-0 | 0-0 |
| 1 | 0 | 0 | 0-- | 0-- |

$$K_1, K_2, K_3$$

$$K_1 = xy_2$$
$$K_2 = 1$$
$$K_3 = y_2 + \bar{x}\bar{y}_1$$

Figure 5.26   JK FF excitation matrix

matrix directly from the state table and state assignment, without generating the $Y$-matrix.

## 5.3.3 STATE ASSIGNMENT

A valid state assignment for a state table $M$ must associate a unique coding in the state variables $y_1, y_2, ..., y_k$ with each state of $M$. Using $S_0$ binary state variables, the total number of codings available is $2^{S_0}$. Hence if $M$ has $n$ states, $S_0$ must be such that $2^{S_0} \geq n$. This condition can be rewritten as $S_0 \geq \lceil \log_2 n \rceil$ where $\lceil x \rceil$ is the smallest integer $\geq x$. The following table specifies the number of state variables required for various values of $n$. Using $S_0$ state variables the encoding of $n$

| Number of States | Number of State Variables Required |
|:---:|:---:|
| 2 | 1 |
| $3 \leq n \leq 4$ | 2 |
| $5 \leq n \leq 8$ | 3 |
| $9 \leq n \leq 16$ | 4 |
| . | . |
| . | . |
| . | . |
| $2^{i-1} + 1 \leq n \leq 2^i$ | $i$ |

states can be done in many ways. The total number of unicode state assignments is

$$\binom{2^{S_0}}{n} n! = 2^{S_0}! / (2^{S_0} - n)! \text{ where } i! = \prod_{j=1}^{i} j,$$

which corresponds to selecting $n$ of the $2^{S_0}$ codes in all ways and assigning them to the $n$ states in all possible ways. However some of these state assignments are effectively identical in that they lead to identical circuit realizations. For example consider the state assignments of Figure 5.27. State Assignment 2 can be derived from state assignment #1 by interchanging the variables $(y_2' = y_1, y_1' = y_2)$. Of course this does not alter the logical complexity of the circuit. Similarly each variable of Assign-

| State Assignment 1 State | $y_1$ | $y_2$ | State Assignment 2 State | $y_1'$ | $y_2'$ | State Assignment 3 State | $y_1''$ | $y_2''$ |
|---|---|---|---|---|---|---|---|---|
| $q_1$ | 0 | 0 | $q_1$ | 0 | 0 | $q_1$ | 1 | 1 |
| $q_2$ | 0 | 1 | $q_2$ | 1 | 0 | $q_2$ | 1 | 0 |
| $q_3$ | 1 | 0 | $q_3$ | 0 | 1 | $q_3$ | 0 | 1 |
| $q_4$ | 1 | 1 | $q_4$ | 1 | 1 | $q_4$ | 0 | 0 |

Figure 5.27   Three 4-state assignments

ment 3 corresponds to the complement of a variable of Assignment 1 ($y_1'' = \bar{y}_1, y_2'' = \bar{y}_2$). Assuming the excitation function for each variable is realized as a separate minimal two level circuit, the complexity will not be altered by complementing a variable. Under these assumptions two state assignments $A$ and $B$ are equivalent if each variable of $A$ is identical to, or the complement of, a variable of $B$. (If we assume that the excitation functions will be considered as a set, and a shared logic two level realization derived, then assignments $A$ and $B$ are equivalent if all variables of $A$ are identical to variables of $B$ or all variables of $A$ are the complements of variables of $B$).

For a unicode assignment with $S_0$ state variables there are $S_0!$ permutations of the variables and $2^{S_0}$ ways of complementing the set of $S_0$ variables. Thus it can be shown[15] that the number of intrinsically different state assignments is

$$\frac{2^{S_0}!/(2^{S_0} - n)!}{S_0!2^{S_0}} = \frac{(2^{S_0} - 1)!}{S_0!(2^{S_0} - n)!}$$

The table of Figure 5.28 shows the number of state assignments for different values of $n$.

| $n$ | $S_0$ | Number of State Assignments |
|---|---|---|
| 4 | 2 | 3 |
| 5 | 3 | 140 |
| 8 | 3 | 840 |
| 9 | 4 | $>10^6$ |

Figure 5.28   Number of state assignments

In general different state assignments will lead to logic realizations which may vary considerably in complexity. Furthermore the determination of the state assignment which leads to the most economical realization will be influenced by the type of memory elements being utilized. Thus for the state table of Figure 5.29, and the state assignments

| | $x$ | | State Assignment A | | State Assignment B | |
|---|---|---|---|---|---|---|
| | 0 | 1 | $y_1$ | $y_2$ | $y_1'$ | $y_2'$ |
| $q_1$ | $q_3$ | $q_2$ | 0 | 0 | 0 | 0 |
| $q_2$ | $q_4$ | $q_1$ | 0 | 1 | 1 | 1 |
| $q_3$ | $q_1$ | $q_3$ | 1 | 0 | 1 | 0 |
| $q_4$ | $q_2$ | $q_3$ | 1 | 1 | 0 | 1 |

Figure 5.29  A state table and two state assignments

$A$ and $B$, if $T$ flip flops are used as memory elements, assignment $A$ leads to the excitation functions

$$T_1 = \bar{x}$$
$$T_2 = x(\bar{y}_1 + y_2)$$

while assignment $B$ leads to the excitation functions

$$T_{1'} = \bar{x} + \bar{y}_1' + y_2'$$
$$T_{2'} = x(\bar{y}_1' + y_2').$$

Therefore assignment $A$ leads to the more economical logic realization. However if $D$-FF's are used as memory elements the respective excitation functions are

$$D_1 = \bar{x}\bar{y}_1 + xy_1$$
$$D_2 = \bar{x}y_2 + x\bar{y}_1\,\bar{y}_2$$
$$D_{1'} = \bar{y}_1' + x\bar{y}_2'$$
$$D_{2'} = \bar{x}y_2' + x\bar{y}_1'\,\bar{y}_2'$$

and assignment $B$ leads to the more economical realization.

For tables of four or fewer states there are only three different (non-equivalent) state assignments, so all assignments can be evaluated before selecting one. However for five states there are 140 different state assignments and this number rises rapidly to more than one million for 9-state tables. It is thus apparent that exhaustive evaluation is not a feasible solution to the problem of deriving a state assignment which leads to the simplest excitation logic. This *state assignment problem* is very difficult and no completely effective solution procedure has been discovered. An algebra of partitions and set systems has been developed which can be effectively utilized to find good state assignments for relatively small tables.[11,12] However this algebra is beyond the scope of this book. We shall consider another state assignment procedure which is based on the following much simpler concept. A state assignment will lead to a reasonably economical realization if in the excitation and $Z$ matrices the entries tend to be clustered into separate large groups of 0's or groups of 1's. Assuming *D-FF* memory elements, if $N(q_i, I_m)$ $= q_j$ and $N(q_i, I_p) = q_k$, if $I_m$ and $I_p$ are adjacent columns and the state variable codings for $q_k$ and $q_j$ are the same in $r$ state variables, then these $r$ excitation functions will have a cluster of two 1's or 0's. Similarly if $N(q_i, I_m) = N(q_j, I_m)$ then if the codings for states $q_i$ and $q_j$ are adjacent (i.e. differ in only one state variable) all excitation functions will have a cluster of two 1's or 0's, and if $Z(q_i, I_m) = Z(q_k, I_m)$, the $Z$-matrix will also have a cluster of two 1's or 0's. Finally, if $N(q_i, I_m) = q_i$ and $N(q_j, I_m) = q_j$, or $N(q_i, I_m) = q_j$ and $N(q_j, I_m) = q_i$, then if states $q_i$ and $q_j$ have adjacent codes, all but one of the excitation maps will have a cluster of two 1's or 0's. Thus from the state table we can derive a weight associated with each pair of states which reflects in a general way, the value of states $q_i$ and $q_j$ being assigned adjacent codings, according to the following procedure, assuming *D-FF* memory elements.

## Procedure 5.3

1) Let $n$ be the number of states and $S_0 = \lceil \log_2 n \rceil$. Associate a weight $w(q_i, q_j)$ with each pair of states. Initially set $w(q_i, q_j) = 0$ for all pair of states.

2) Evaluate all of the state table columns $I_m$. If $N(q_i, I_m) = N(q_j, I_m)$ add $S_0$ to $w(q_i, q_j)$.

3) If $z_k(q_i, I_m) = z_k(q_j, I_m)$ add 1 to $w(q_i, q_j)$.

4) Evaluate all of the table rows. If $I_m$ and $I_p$ are adjacent input

columns and $N(q_i,I_m) = q_j$, $N(q_i,I_p) = q_k$ add 1 to $w(q_j,q_k)$.*

5) If $N(q_i,I_m) = q_i$ and $N(q_j,I_m) = q_j$, or if $N(q_i,I_m) = q_j$ and $N(q_j,I_m) = q_i$, add $S_0 - 1$ to $w(q_i,q_j)$.

6) A state assignment $A$ has a weight equal to the sum of the weights of all state pairs which have adjacent codings in $A$. State assignments with high weights will tend to have relatively economical realizations. A state assignment with a relatively high weight can be derived in an ad hoc manner, by assigning as many as possible of those state pairs $q_i,q_j$ with large weights $w(q_i,q_j)$ to adjacent codings.

**Example 5.7**

For the state table of Figure 5.30(a), $S_0 = 2$, and the weights associated

$$x_1 x_2$$

| | 00 | 01 | 11 | 10 |
|---|---|---|---|---|
| $q_1$ | $q_3,0$ | $q_1,1$ | $q_2,0$ | $q_1,0$ |
| $q_2$ | $q_2,0$ | $q_4,1$ | $q_2,1$ | $q_3,0$ |
| $q_3$ | $q_4,0$ | $q_3,0$ | $q_3,0$ | $q_3,1$ |
| $q_4$ | $q_4,1$ | $q_3,0$ | $q_1,1$ | $q_4,1$ |

(a)

| State Pair | Weights from (2) | Weights from (3) | Weights from (4) | Weights from (5) | Total Weight |
|---|---|---|---|---|---|
| $q_1 q_2$ | 2 | 3 | 2 | 0 | 7 |
| $q_1 q_3$ | 0 | 2 | 3 | 2 | 7 |
| $q_1 q_4$ | 0 | 0 | 1 | 1 | 2 |
| $q_2 q_3$ | 2 | 1 | 2 | 1 | 6 |
| $q_2 q_4$ | 0 | 1 | 2 | 1 | 4 |
| $q_3 q_4$ | 4 | 2 | 3 | 1 | 10 |

(b)

Figure 5.30   (a) A state table (b) the state pair weights

*This condition is less important than that of (2) or (3), since they require adjacency for *any* clustering effect while the clustering effect of this condition is proportional to the number of variables in which the codings differ and does not require adjacency.

with each pair of states are as shown in the table of Figure 5.30(b) in which the columns indicate the contributions to the total weight (last column) from each of steps (2)–(5) of Procedure 5.3.

State assignment 1 has total weight of 28 and leads to a two level logic realization (without sharing) with $D$-$FF$'s and having 47 gate inputs. Similarly state assignment 2 has total weight of 25 and its realization requires 55 gate inputs while state assignment 3 has weight 19 and requires 63 gate inputs. This demonstrates the general correlation between the weight of a state assignment and the cost of the corresponding realization.

| State Assignment 1 | | | State Assignment 2 | | | State Assignment 3 | | |
|---|---|---|---|---|---|---|---|---|
| State | $y_{11}$ | $y_{21}$ | State | $y_{12}$ | $y_{22}$ | State | $y_{13}$ | $y_{23}$ |
| $q_1$ | 0 | 0 | $q_1$ | 0 | 0 | $q_1$ | 0 | 0 |
| $q_2$ | 0 | 1 | $q_2$ | 0 | 1 | $q_2$ | 1 | 1 |
| $q_3$ | 1 | 0 | $q_3$ | 1 | 1 | $q_3$ | 0 | 1 |
| $q_4$ | 1 | 1 | $q_4$ | 1 | 0 | $q_4$ | 1 | 0 |

Figure 5.31   Three state assignments

**Multicode State Assignments**

For an $n$-state table if $n \neq 2^{S_0}$ a unicode state assignment will result in a $Y$-matrix with $2^{S_0} - n$ rows corresponding to codes which are not assigned to any state and hence these rows have don't care entries. It is possible to use some of these spare codes by assigning multiple codes to some states, so long as no coding is assigned to more than one state. Such a coding is referred to as a *multicode state assignment*.

In the $Y$-matrix for such an assignment, the next state entries corresponding to a state $q_i$ can be filled with any of the codings assigned to $q_i$. *All rows corresponding to codings for state $q_i$ must be filled* with entries consistent with the codings for the next state entries of state $q_i$. In general, a multicode assignment will have fewer don't cares in the $Y$-matrix than a unicode assignment. However, it may be possible that the don't cares are distributed in such a way that the logic is simplified[5]. The previously mentioned set system algebra may be used to generate such state assignments. However these state assignments will not be further considered in this book.

### 5.3.4 REDUCTION OF STATE TABLES

Given a state table $M$ it is sometimes possible to find another state table $M'$ whose behavior is essentially identical to that of $M$. If $M'$ has fewer states than $M$, then we say the number of states of $M$ can be *reduced*. Reducing the number of states in a state table will often enable the simplification of the logic of a circuit realizing the state table. For example consider the two state tables shown in Figure 5.32.

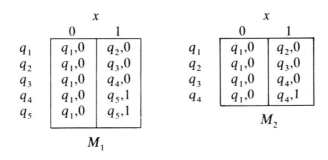

Figure 5.32  Two equivalent state tables

For Table $M_1$ it is obvious that the output sequence for any input sequence $I$ and initial state $q_4$ will be the same as the output sequence for $I$ and initial state $q_5$. Therefore the output sequence for any input sequence will be unchanged if all next state entries equal to $q_5$ are changed to $q_4$. Furthermore $q_5$ does not appear as a next state entry after this modification, and $q_5$ and $q_4$ have identical next state/output entries in both columns of the table. Therefore $q_5$ is equivalent to $q_4$ as an initial state and state $q_5$ can be eliminated from the table thus resulting in the table $M_2$ which requires two state variables while table $M_1$ requires three state variables. In this section we shall consider the problem of reducing the number of states in *completely specified tables*.

Formally state $q'$ of table $M'$ *covers* state $q$ of table $M$ if the output sequences of $M'$ in initial state $q'$ and $M$ in initial state $q$ are identical *for any input sequence*. Table $M'$ *covers* table $M$ if every state of $M$ is covered by some state of $M'$. If $M'$ covers $M$ and $M$ covers $M'$ (as is the case for completely specified tables) then $M'$ is *equivalent* to $M$. The *state table reduction (or minimization) problem* is: for a given table $M$ find a table $M'$ which covers $M$, such that for any other table $M''$ which covers $M$, the number of states in $M'$ does not exceed the number of states in $M''$.

If a table $M$ can be covered by another table $M'$ containing fewer states than $M$, the same state of $M'$ must cover more than one state of $M$. If a set of states of $M$ can be covered by the same state of $M'$ the states in this set are said to be *equivalent*. Determining equivalent sets of states of a state table is essential to the state minimization problem. In order for a set of states $Q = \{q_1, q_2, ..., q_r\}$ of a table $M$ to be equivalent they must generate the same output sequence for any input sequence. In order for this to be true for all input sequences of length one it is necessary that for any input $I_k$, $Z(q_1, I_k) = Z(q_2, I_k) = \cdots = Z(q_r, I_k)$. A set of states $Q$ which has this property will be called *1-equivalent*. In order for all states of $Q$ to generate the same output sequence for all input sequences of length two, $Q$ must be 1-equivalent and for any input $I_k$, the set of states $\{N(q_1, I_k), N(q_2, I_k), \cdots, N(q_r, I_k)\}$ must be equivalent. Thus a pair of states $(q_1, q_2)$ will not be equivalent if and only if $q_1$ and $q_2$ are not 1-equivalent or for some input $I_k$, $N(q_1, I_k)$ and $N(q_2, I_k)$ are not equivalent. We can effectively find all pairs of states which are not equivalent and hence all equivalent state pairs, by the following procedure.

**Procedure 5.4**

(1) Form a list $L$ consisting of all state pairs which are not 1-equivalent.

(2) Add to $L$ any pair of states $(q_i, q_j)$ if for some input $I_k$ the state pair $N(q_i, I_k), N(q_j, I_k))$ is in $L$ (i.e., $N(q_i, I_k)$ and $N(q_j, I_k)$ are a previously determined nonequivalent pair of states). The state pair $(q_i, q_j)$ is said to *imply* the state pair $( N(q_i, I_k), N(q_j, I_k))$ if $N(q_i, I_k) \neq N(q_j, I_k)$.

(3) Repeat (2) until no new pairs can be added to $L$. $L$ contains all nonequivalent state pairs. All state pairs not in $L$ are equivalent.

This procedure for finding equivalent pairs of states of a table can be tabularized using a *pair chart*. For an $n$-state table this chart is shown in Figure 5.33. For any pair of states $(q_i, q_j)$, $i < j$, there is a corresponding box in the pair chart in column $i$, row $j$. If states $q_i$ and $q_j$ are not 1-equivalent we place a cross ($\times$) in this box. Otherwise the box is filled with entries $q_m q_n$ denoting the pairs of states implied by $(q_i, q_j)$ for each input. If $(q_i, q_j)$ implies no state pairs (i.e., $N(q_i, I_k) = N(q_j, I_k)$ for all inputs $I_k$) a check ($\checkmark$) is placed in that box. After filling all boxes in this manner, we then place $\times$'s in every box which contains a state pair $q_m q_n$ if the box corresponding to $(q_m, q_n)$ has an $\times$. This is repeated until no more $\times$'s can be entered in the pair chart. All nonequivalent pairs now have $\times$'s in their respective boxes.

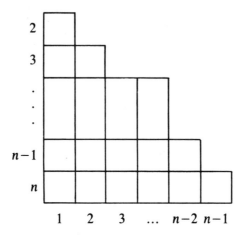

**Figure 5.33   A pair chart**

## Example 5.8

Consider the state table of Figure 5.34. The set of state pairs which are not 1-equivalent is $\{(q_1,q_4), (q_2,q_4), (q_3,q_4), (q_4,q_5), (q_4,q_6)\}$. Therefore the initial pair chart is as shown in Figure 5.35(a). Note that state pair $(q_1 q_3)$ implies both $q_3 q_5$ (in column 0) and $q_2 q_6$ (in column 1). From this pair chart it is seen that the state pairs $\{(q_1,q_2), (q_1,q_6), (q_2,q_3), (q_2,q_5), (q_3,q_6), (q_5,q_6)\}$ imply state pairs which are not equiv-

$$x$$

| | 0 | 1 |
|---|---|---|
| $q_1$ | $q_3,0$ | $q_2,0$ |
| $q_2$ | $q_3,0$ | $q_4,0$ |
| $q_3$ | $q_5,0$ | $q_6,0$ |
| $q_4$ | $q_5,1$ | $q_1,1$ |
| $q_5$ | $q_1,0$ | $q_2,0$ |
| $q_6$ | $q_5,0$ | $q_4,0$ |

**Figure 5.34   A state table**

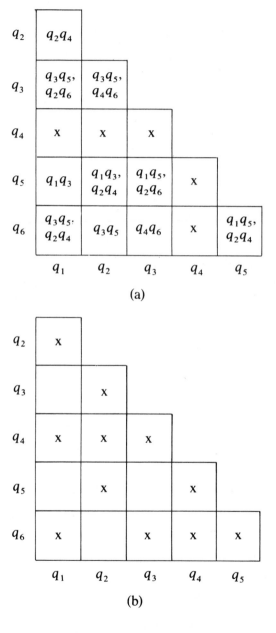

Figure 5.35   Pair chart generation process

alent. The $\times$'s of the pair chart are then as shown in Figure 5.35(b). Now no remaining state pair implies a state pair which is not equivalent and hence the set of state pairs $\{(q_1,q_3), (q_1,q_5), (q_2,q_6), (q_3,q_5)\}$ are equivalent.

From the pair chart, sets of equivalent states can easily be determined. This is because, for completely specified tables, all states which are equivalent to a state $q_i$ are also equivalent to each other. To prove this we first prove the following Lemma.

## Lemma 5.1

For a completely specified table $M$, if $q_i \equiv q_k$ ($q_i$ is equivalent to $q_k$) and $q_j \equiv q_k$ then $q_i \equiv q_j$.

## Proof:

Assume $q_i \not\equiv q_j$. Then there exists an input sequence $\mathbf{I}$ such that $M$ in initial state $q_i$ generates the output sequence $\mathbf{Z}$ in response to $\mathbf{I}$ and in initial state $q_j$ generates the output sequence $\mathbf{Z}'$ in response to $\mathbf{I}$ and $\mathbf{Z} \neq \mathbf{Z}'$. Let us denote the output sequence generated by applying $\mathbf{I}$ to initial state $q_k$ as $\mathbf{Z}^*$. Since $q_i \equiv q_k, \mathbf{Z} = \mathbf{Z}^*$, and since $q_j \equiv q_k$, $\mathbf{Z}' = \mathbf{Z}^*$. Therefore $\mathbf{Z} = \mathbf{Z}'$. Contradiction.

## Theorem 5.3:

For a completely specified table if states $q_1, q_2, ..., q_p$ are all equivalent to $q_k$ then the set of states $Q = \{q_1, q_2, ..., q_p, q_k)\}$ is equivalent.

## Proof

Suppose the set of states $Q$ is not equivalent. Then there must be some pair of states $q_i$ and $q_j$ in $Q$ such that $q_i \not\equiv q_j$. However $q_i \equiv q_k$ and $q_j \equiv q_k$. Therefore, by Lemma 5.1, $q_i \equiv q_j$. Contradiction.

The following procedure can be used to derive a collection of equivalent sets of states from the pair chart.

## Procedure 5.5

(1) Let $i = 1$
(2) Create a set $B_1$ consisting of all states equivalent to $q_1$.
(3) Increase $i$ by 1
(4) Let $q_j$ be a state not contained in any set $B_k$ generated so far,

such that for any other such state $q_m$, $j < m$. Create a set $B_i$ consisting of all states equivalent to $q_j$.
(5) Repeat (3) and (4) until all states appear in some block $B_i$.

It follows from Lemma 5.1 that the blocks, $B_i$, generated by Procedure 5.5 will be disjoint, and hence define a partition $\pi_E$ on the set of states.

## Example 5.9
For the pair chart of Figure 5.35(b) states $q_3$ and $q_5$ are equivalent to $q_1$, and generate the set $B_1 = \{q_1, q_3, q_5\}$. State $q_2$ is not in $B_1$ and $q_6$ is equivalent to $q_2$. Thus $B_2 = \{q_2, q_6\}$. State $q_4$ is not in $B_1$ or $B_2$ and $q_4$ is not equivalent to any other state. Thus $B_3 = \{q_4\}$. Thus Procedure 5.5 applied to this pair chart generates the partition $\pi_E = \{(q_1, q_3, q_5), (q_2, q_6), q_4\}$.

## Lemma 5.2
For a completely specified table, if $q_i$ and $q_j$ are in different blocks of $\pi_E$ then $q_i \not\equiv q_j$.

## Proof
Suppose $q_i$ and $q_j$ are in different blocks of $\pi_E$ and $q_i \equiv q_j$. Without loss of generality assume $q_i$ is in a block $B_m$ and $q_j$ is in a block $B_n$ and $m < n$. If $B_m$ was defined as all states equivalent to $q_m$, then $q_m \equiv q_i$. By assumption $q_i \equiv q_j$. Therefore, by Lemma 5.1, $q_j \equiv q_m$ and $q_j$ is in $B_m$. Contradiction.

As a consequence of this result the partition $\pi_E$ can be used to generate a minimal reduced table $M'$ equivalent to the given table $M$, as follows.

## Procedure 5.6
(1) The inputs of $M'$ are identical to the inputs of $M$. The states of $M'$ correspond to sets of equivalent states of $M$ as defined by the blocks $B_i$ of $\pi_E$.
(2) The next state and output entries of $M'$ are defined as follows. If the present state is $B_i = \{q_{i_1}, ..., q_{ik}\}$ and the input is $I_j$, let $B_{ij} = \{N(q_{i_1}, I_j), N(q_{i2}, I_j)...N(q_{ik}, I_j)\}$. Then, since all states in set $B_{ij}$ are equivalent, for some (unique) block $B_k$ of $\pi_E$, $B_{ij} \subseteq B_k$. Define the next state entry $N(B_i, I_j)$ of $M'$ to be $B_k$, and the output entry $Z(B_i, I_j) = Z(q_{i1}, I_j) = Z(q_{i2}, I_j) = ... = Z(q_{ik}, I_j)$.

## Example 5.10

For the table of Figure 5.34, $\pi_E = \{(q_1,q_3,q_5), (q_2,q_6), (q_4)\}$. The reduced machine $M'$ defined by $\pi_E$ is shown in Figure 5.36.

$$x$$

| | | 0 | 1 |
|---|---|---|---|
| $\{q_1,q_3,q_5\}$ | $B_1$ | $B_1,0$ | $B_2,0$ |
| $\{q_2,q_6\}$ | $B_2$ | $B_1,0$ | $B_3,0$ |
| $\{q_4\}$ | $B_3$ | $B_1,1$ | $B_1,1$ |

**Figure 5.36  Reduced state table**

The next state entry $N(B_1,1)$ is computed as follows. Since $B_1 = \{q_1,q_3,q_5\}$, $B_{11} = \{N(q_1,1), N(q_3,1), N(q_5,1)\} = \{q_2,q_6\} = B_2$. Therefore $N(B_1,1) = B_2$. The output entry $Z(B_1,1)$ is equal to $Z(q_1,1) = Z(q_3,1) = Z(q_5,1) = 0$.

Furthermore the machine defined by Procedure 5.6 is the unique minimal state machine which is equivalent to $M$.

## Theorem 5.4

The machine $M'$ generated by $\pi_E$ is the unique minimal state machine which is equivalent to the given machine $M$.

## Proof

Suppose there is another machine $M^*$ which is equivalent to $M$ and is different from $M'$. Every state of $M^*$ is equivalent to a set of states of $M$. Suppose some state $q_i$ of $M^*$ is equivalent to a set of states $\{q_{il}, ..., q_{ik}\}$ of $M$, and no state of $M'$ is equivalent to this set of states. Then by Lemma 5.2 for two of these states $q_{im}$ and $q_{in}$, $q_{im} \not\equiv q_{in}$. Therefore $M^*$ is not equivalent to $M$.

The following example demonstrates the state minimization process for completely specified state tables in its entirety.

## Example 5.11

For the state table of Figure 5.37(a), the equivalence pair chart is shown in Figure 5.37(b). The equivalence partition $\pi_E = \{(q_1,q_6,q_7),$

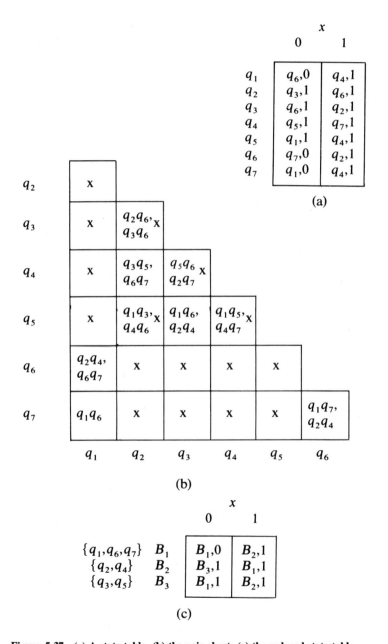

Figure 5.37   (a) A state table, (b) the pair chart, (c) the reduced state table

$(q_2,q_4), (q_3,q_5)\}$ defines the minimal state equivalent table shown in Figure 5.37(c).

For incompletely specified state tables the state minimization problem is much more difficult. Since output sequences may not be completely specified the definition of equivalence is modified so that states $q_i$ and $q_j$ of a table $M$ are equivalent if the output sequence generated in response to any input $I_k$ from initial states $q_i$ or $q_j$ are identical at all times when both such outputs are specified. Lemma 5.1 is not valid in this case, and as a consequence we cannot define a partition which satisfies Lemma 5.2 and which can be used to define a minimal state equivalent machine. The state minimization problem for incompletely specified state tables is considered in great detail in more advanced books[5] to which the interested reader is referred.

## 5.4 ANALYSIS OF SEQUENTIAL CIRCUITS

The steps in the analysis of sequential circuits can be represented by the block diagram of Figure 5.38. As with the problem of transforming

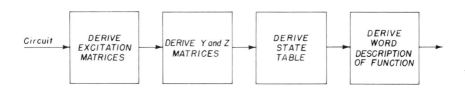

Figure 5.38    Analysis of sequential circuits

a word description into a state table, the process of translating a table into a word description is not well-defined. However, in many cases it can be accomplished in a relatively straightforward manner as illustrated in the following example.

### Example 5.12
The output of the state table of Figure 5.39 is 1 only if the present input is 1 and the state is $q_2$. However, the state is $q_2$ only if the previous input was 0 and the previous state was $q_1$ or $q_2$. The previous state was $q_1$ or $q_2$ only if the next previous input was 0. Therefore,

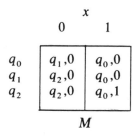

Figure 5.39   A finite state table

the output of $M$ is 1 if and only if the present input is 1 and the previous two inputs were 0 (assuming $q_0$ as an initial state).

The process of generating a state table from a circuit is the reverse of deriving a circuit from a state table. The first step is to construct the memory element excitation matrices and a map of the outputs $z$ as functions of $y$ and $x$. This problem is identical to the analysis of a combinational circuit with inputs $x$ and $y$, and outputs $z$ and $E_i$, where $E_i$ are the memory excitation inputs. (The excitation matrices will always be completely specified). The flip-flop state tables (Figures 5.7–5.10) can be used to convert the excitation matrix of the $Z$-matrix to a $Y$-$Z$-matrix. By associating a state with each state variable coding the $Y$-$Z$-matrix is transformed to a state table. This state table will always have $2^k$ states, where $k$ is the number of flip-flops in the circuit, and may be reducible.

## Example 5.13

For the circuit of Figure 5.40, the $FF$ excitation functions are

$$J_1 = \bar{x}y_2 + x\bar{y}_2 \qquad\qquad J_2 = x + y_1$$
$$K_1 = \bar{x} \qquad\qquad K_2 = x + \bar{y}_1$$

and the output is $z = \bar{x}y_1 y_2$.

Plotting these expressions on a Karnaugh map generates the excitation matrix and $Z$ map of Figure 5.41.

Using the state table for $JK$ flip flops of Figure 5.8(c), the $Y$-$Z$-matrix (Figure 5.42(a)) is derived from these matrices. The entry in row (10) and column 1 has $J_1 = 1$ and $K_1 = 0$ so $Y_1 = 1$, and $J_2 = 1, K_2 = 1$

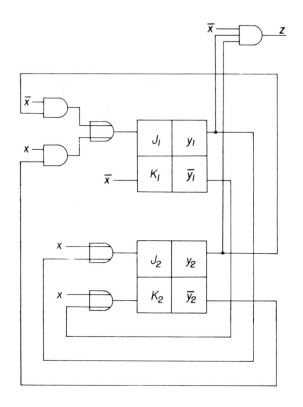

**Figure 5.40  A sequential circuit**

|  | $x$ | | |  | $x$ | |
|---|---|---|---|---|---|---|
| $y_1\ y_2$ | 0 | 1 | $y_1\ y_2$ | 0 | 1 |
| 0 0 | 01, 01 | 10, 11 | 0 0 | 0 | 0 |
| 0 1 | 11, 01 | 00, 11 | 0 1 | 0 | 0 |
| 1 1 | 11, 10 | 00, 11 | 1 1 | 1 | 0 |
| 1 0 | 01, 10 | 10, 11 | 1 0 | 0 | 0 |

$$J_1 K_1, J_2 K_2 \qquad\qquad z$$

**Figure 5.41  Excitation matrix and Z matrix**

| $y_1 \, y_2$ | $x$ 0 | 1 | | $x$ 0 | 1 |
|---|---|---|---|---|---|
| 0 0 | 00, 0 | 11, 0 | $q_1$ | $q_1,0$ | $q_3,0$ |
| 0 1 | 10, 0 | 00, 0 | $q_2$ | $q_4,0$ | $q_1,0$ |
| 1 1 | 01, 1 | 10, 0 | $q_3$ | $q_2,1$ | $q_4,0$ |
| 1 0 | 01, 0 | 11, 0 | $q_4$ | $q_2,0$ | $q_3,0$ |
| | (a) | | | (b) | |

Figure 5.42    Y-Z matrix and state table

so variable $y_2$ changes state and $Y_2 = 1$ since in row (10), $y_2 = 0$.

Associating the states $q_1, q_2, q_3, q_4$ with state variable codings 00, 01, 11, 10, respectively, we obtain the state table of Figure 5.42(b).

## 5.5 SEQUENTIAL MACHINE EXPERIMENTS

The behavior of finite-state machines can sometimes be analyzed by experiments, in which an input sequence (or set of input sequences) is applied to the circuit, the output sequence (or sequences) are observed and from this some information about the machine can be inferred without knowledge of the exact circuit but only the state table of the sequential function. One such experiment is a *synchronizing sequence*. This is an input sequence, which when applied to the machine results in a unique final state independent of the initial state. Thus for the table of Figure 5.43 the input sequence 010 applied to any initial state, will result in a unique final state, $q_4$.

| | $x$ 0 | 1 |
|---|---|---|
| $q_1$ | $q_4,0$ | $q_3,1$ |
| $q_2$ | $q_4,1$ | $q_1,1$ |
| $q_3$ | $q_2,0$ | $q_1,0$ |
| $q_4$ | $q_3,1$ | $q_2,0$ |

Figure 5.43    A finite state table

A synchronizing sequence can be derived from a *successor tree*. The initial (top) node $Q$ of this tree represents the set of all possible initial states. For each input $I_j$, there is a branch from $Q$ to a node $N_j$ where $N_j = \{q_i | q_i = N(q_k, I_j),$ for some $q_k \in Q\}$. Branches are similarly defined from a node $N_j$ to $N_{jp} = \{q_i | q_i = N(q_k, I_p),\ q_k \in N_j\}$. If any node $N_I$ represents a single state $q_m$, then $\mathbf{I}$ is a synchronizing sequence which always results in final state $q_m$. For the table of Figure 5.43, the successor tree is as shown in Figure 5.44 from which it is concluded that 010

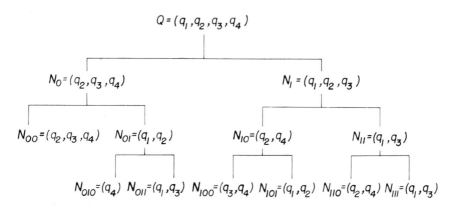

**Figure 5.44   Successor tree**

is a synchronizing sequence since $N_{010} = \{q_4\}$. Note that no branches were generated from $N_{00}$ since $N_{00}$ is identical to one of its predecessors, $N_0$, and hence the successors of $N_{00}$ will be identical to the successors of $N_0$.

Some state tables have no synchronizing sequence. For example the table of Figure 5.45(a) generates the successor tree of Figure 5.45(b).

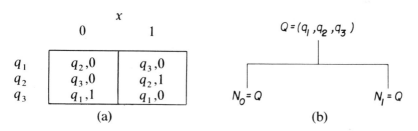

|       | $x$ | |
|-------|-----------|-----------|
|       | 0         | 1         |
| $q_1$ | $q_2,0$   | $q_3,0$   |
| $q_2$ | $q_3,0$   | $q_2,1$   |
| $q_3$ | $q_1,1$   | $q_1,0$   |
|       | (a)       |           |

**Figure 5.45   (a) State table and (b) successor tree**

Since no node of this tree represents a single node, the table has no synchronizing sequence.

A *distinguishing sequence* is an input sequence which results in a different output sequence for every different initial state. Hence knowledge of the input sequence *and* output sequence is sufficient to determine both the initial and final states. A *homing sequence* is an input sequence which results in a different output sequence for every different final state which can be reached by applying the input sequence to any initial state. Thus knowledge of the input sequence and output sequence is sufficient to determine the final state but not necessarily the initial state of the machine. For the table of Figure 5.46, the input sequence 010

$$x$$

|       | 0         | 1         |
|-------|-----------|-----------|
| $q_1$ | $q_2,0$   | $q_2,1$   |
| $q_2$ | $q_3,0$   | $q_4,0$   |
| $q_3$ | $q_4,0$   | $q_1,0$   |
| $q_4$ | $q_1,1$   | $q_2,1$   |

**Figure 5.46    A finite state table**

is a distinguishing sequence since it results in an output sequence of 001 from initial state $q_1$, 000 from initial state $q_2$, 010 from initial state $q_3$, and 110 from initial state $q_4$. The input sequence 11 is a homing sequence since it results in a final state $q_4$ with output sequence 10 if the initial state is $q_1$ or $q_4$, or a final state $q_2$ with output sequence 01 if the initial state is $q_2$ or $q_3$.

Trees similar to the previously defined successor tree, can be used to derive distinguishing sequences and homing sequences. Not all tables have distinguishing sequences but every reduced table has a homing sequence. (Problem 5.18) Other experiments have been defined which enable one machine $M$ to be distinguished from any other machine $M'$, with the same number of inputs, states, and outputs, if $M$ is reduced and *strongly connected*.* A good reference on the subject of sequential machine experiments is a book by Gill[6].

*A machine $M$ is strongly connected if for any ordered pair of states $q_i, q_j$ of $M$, there exists an input sequence **I** which causes $M$ to go from $q_i$ to $q_j$.

## 5.6 LIMITATION OF FINITE STATE MACHINES

There exist computations which cannot be performed by finite state machines. These are computations which may require an arbitrarily large amount of memory. Consider the design of a machine whose input is a sequence of 0's and 1's. Starting in an initial state $q_0$ at time $t_0$ we want the output $z$ at time $t \geq t_0$ to be 1 if the number of 0-inputs from $t_0$ to $t$ is equal to the number of 1-inputs from $t_0$ to $t$, and $z$ should be 0 if the number of 0-inputs is not equal to the number of 1-inputs from $t_0$ to $t$. It is impossible to design a finite state sequential machine which can perform this computation. Informally this is because such a machine must remember at each instant of time $t$, the difference between the number of 0- and 1-inputs from $t_0$ to $t$. But this quantity may be arbitrarily large and hence such a machine must have a potentially unlimited amount of memory. Formally suppose machine $M$ has $k$ states and $M$ performs the computation specified. If we apply the input sequence $I = \underbrace{00...0}_{r \text{ times}} \underbrace{11...1}_{r \text{ times}}$ to $M$ in initial state $q_0$ the last output should be 1. Let us denote the state after the $i$-th input as $q_i$. Suppose $r > k$. Then $q_r$ must be the same as $q_j$ for some $j < r$ since the machine has $k < r$ states and hence must enter a state cycle in which it cycles through a set of states periodically when given $r > k$ consecutive 0-inputs. Furthermore the input sequence $\underbrace{11...1}_{r \text{ times}}$ applied to state $q_r$ of $M$ results in a 1-output. Therefore the input sequence $I' = \underbrace{00...0}_{j \text{ times}} \underbrace{11...1}_{r > j \text{ times}}$ applied to $M$ in initial state $q_0$ results in the last output equal to 1. But this is incorrect since at that time the number of 0-inputs and 1-inputs are unequal. Hence $M$ does not perform the desired computation.

Another type of function which cannot be computed by a finite state sequential machine is sequential multiplication. Such a machine would accept two binary number strings sequentially, low order bits first (i.e. $(a_0 b_0)(a_1 b_1)...(a_i b_i))$ and generate the product $A \cdot B = (\Sigma a_i 2^i) \cdot (\Sigma b_i 2^i)$ as a sequential pattern of binary numbers, low order bit first. The first output of such a circuit would be equal to $a_0 \cdot b_0$. The second output would be $a_1 \cdot b_0 \oplus a_0 \cdot b_1$ and would result in a carry of 1 if $a_1 = a_0 = b_1 = b_0 = 1$. The 3-rd output is equal to $a_2 \cdot b_0 \oplus a_1 \cdot b_1 \oplus a_0 \cdot b_2 \oplus C$ where $C$ is the carry in, and results in a carry out of 1 or 2. Continuing in this manner we observe that the $i$-th *bit* can result in a carry proportional to $\log_2 i$. Since the carry must be remembered, an unlimited amount of memory is required to compute the product of two arbitrarily long binary numbers. Of course for any specific value of $k$ we can design a finite state sequential machine to multiply two $k$-bit numbers sequentially. However such a circuit may vary greatly for different values of $k$. In the next chapter we shall show how a multiplier circuit can be designed on a *system level* in such a way that it is easily modified from a $k$-bit to a $k'$-bit multiplier.

The *Turing machine* is a machine model which is a generalization of a finite state sequential machine. It can perform many computations which cannot be performed by a finite state sequential machine including the two illustrations previously considered. A Turing machine has a control unit with a finite number of states. The inputs to the machine are written on an arbitrarily long tape and the control can read from the tape and write on the tape and move in both directions on the tape. The operation of the Turing machine is as follows: Let the machine be initially in some state $q_0$ at some position $P$ on the tape. On reading the input symbol $x_i$ on the tape, the control unit overwrites $x_i$ with some symbol $x_j$, changes state to $q_j$ and moves either to the right or the left of the symbol just processed. Since the control unit has a finite number of states, a Turing machine can be described by a table similar to the state table. For every combination of input symbol (on the tape) and internal state of the control unit, the table specifies the next state, the symbol to be written on the tape and the direction of shift.

Turing machines are used primarily for the study of computability and computational complexity. Such machines represent mathematical models and are not used in digital circuit design, and therefore we shall not discuss them further.

## 5.7 SEQUENTIAL CIRCUIT DESIGN WITHOUT STATE TABLES

Certain types of sequential circuits, which have a high degree of modularity (i.e. uniform structure) can be designed without using state tables. For example a set of flip-flops can be thought of as a *register*. Some sequential circuits involving operations on registers in which similar logical computations are performed on each *FF* of the register, can be designed directly from a word description of the circuit without using a state table. For instance consider the design of a circulating *shift register* which consists of $k$ flip-flops arranged in a linear cascade, which operates so that at each clock pulse, the contents of each *FF* in the register is shifted to the neighboring *FF* to the right, and the contents of the rightmost *FF* is shifted into the leftmost *FF*.

Assuming the state of the *FF*'s in the register are denoted by $y_1, y_2, ..., y_k$ (from left to right) then all flip-flops of the register have next state equations of similar form

$$Y_i = y_{i-1} \qquad 1 < i \leq k$$
$$Y_1 = y_k$$

The simple circuit of Figure 5.47 using $k$ master-slave flip-flops realizes the desired function. Note that the use of master-slave flip-flops ensures that the contents of the register will be shifted only one position during each clock pulse.

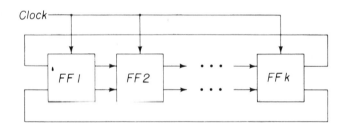

Figure 5.47   Shift register circuit

This type of realization is easily generalized to include the use of *control inputs* to enable or disable the flow of information. For example consider the design of a shift register in which the contents of the register is shifted to the right if control input $C_1 = 1$, and is shifted to the left if $C_1 = 0$. If the state of the *FF*'s in the register are denoted by $y_1, y_2, ..., y_k$ (from left to right) then all flip-flops of the register have similar next state equations of the form

$$Y_i = C_1 \cdot y_{i-1} + \bar{C}_1 y_{i+1} \qquad 1 < y < k$$
$$Y_k = C_1 \cdot y_{k-1} + \bar{C}_1 y_1$$
$$Y_1 = C_1 \cdot y_k + \bar{C}_1 y_2$$

and assuming *SR* flip-flops (master-slave *FF*'s are again required) the excitation functions are of the form

$$
\left.
\begin{aligned}
S_i &= C_1 \cdot y_{i-1} + \bar{C}_1\, y_{i+1} \\
R_i &= C_1 \cdot \bar{y}_{i-1} + \bar{C}_1\, \bar{y}_{i+1}
\end{aligned}
\right\} \quad 1 < i < k
$$

$$
S_1 = C_1\, y_k + \bar{C}_1\, y_2
$$
$$
R_1 = C_1\, \bar{y}_k + \bar{C}_1\, \bar{y}_2
$$
$$
S_k = C_1\, y_{k-1} + \bar{C}_1\, y_1
$$
$$
R_k = C_1\, \bar{y}_{k-1} + \bar{C}_1\, \bar{y}_1
$$

Such a circuit can be realized in the basic form of Figure 5.48 where the modular combinational logic realizes the set of *FF* excitation functions previously defined.

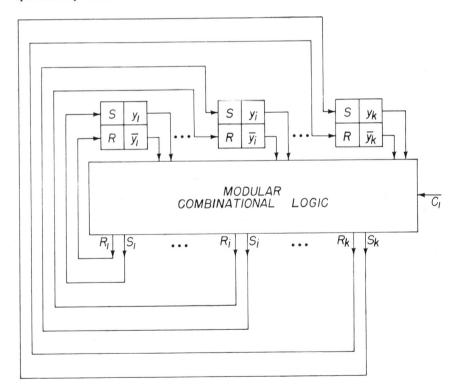

**Figure 5.48   Controlled shift register circuit**

Another common sequential circuit which has a regular design is a *modulo $2^k$ counter*. Such a circuit can be described by a sequential machine with $2^k$ states. The state behavior of the machine is as follows. Starting in an initial state $q_0$, each 1-input received causes the machine to go from present state $q_i$ to next state $q_{i+1}$ for $0 \leq i < 2^k - 1$. After the $2^k$th 1-input, the machine returns to initial state $q_0$. For $k = 3$, the sequential machine is described by the state table of Figure 5.49.*

| | $x$ | | | $y_3$ | $y_2$ | $y_1$ |
|---|---|---|---|---|---|---|
| | 0 | 1 | | | | |
| $q_0$ | $q_0$ | $q_1$ | | 0 | 0 | 0 |
| $q_1$ | $q_1$ | $q_2$ | | 0 | 0 | 1 |
| $q_2$ | $q_2$ | $q_3$ | | 0 | 1 | 0 |
| $q_3$ | $q_3$ | $q_4$ | | 0 | 1 | 1 |
| $q_4$ | $q_4$ | $q_5$ | | 1 | 0 | 0 |
| $q_5$ | $q_5$ | $q_6$ | | 1 | 0 | 1 |
| $q_6$ | $q_6$ | $q_7$ | | 1 | 1 | 0 |
| $q_7$ | $q_7$ | $q_0$ | | 1 | 1 | 1 |

**Figure 5.49    A mod 8 counter**

The realization of a modulo $2^k$ counter requires $k$ state variables. Let us consider a state assignment of the type shown in Figure 5.49 which associates the coding corresponding to the binary encoding of $i$ with state $q_i$, $0 \leq i \leq 2^k - 1$. Using $T$ FF's (of a master-slave variety) as memory elements, and a state assignment of this type, a modulo $2^k$ counter consists of $k$ flip-flops with excitations defined by

$$T_i = x \cdot \prod_{j=1}^{i-1} y_j \qquad \text{for } 1 < i \leq k$$

$$T_1 = x$$

*We may also define a counter which cycles through its states every $2^k$ clock pulses. Such a circuit has only a clock input and is defined by a one-column state table, that column being as shown in column $x = 1$ of the table of Figure 5.49.

Thus this state assignment leads to the modular realization of Figure 5.50. Since a *T-FF* can be realized from an *SR-FF* by the circuit of Figure 5.51, a counter can be designed using *SR-FF*'s, by replacing each *T-FF* in the circuit of Figure 5.50, by the equivalent circuit of Figure 5.51.

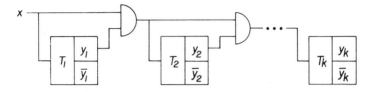

Figure 5.50    Modulo $2^k$ counter circuit

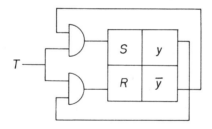

Figure 5.51    Realization of *T* flip-flop

Complex digital systems (e.g. digital multipliers) can frequently be designed from basic building blocks including registers, and counters, sequential circuits which act as controllers, and modular combinational logic circuits. In the next chapter we shall consider such *system level design* of digital systems.

## SOURCES

Several different models of sequential machines have been formulated by Huffman[14], Moore[18] and Mealy[16]. Cadden[3] demonstrated the equivalence of these models. Digital circuit realizations of flip-flops can be found in Millman and Taub[17]. McCluskey and Unger[15] determined the number of different state assignments for an $n$-state table. The state assignment procedure is due to Armstrong[1,2]. A similar procedure has been developed by Dolotta and McCluskey[4]. Another approach to the state assignment problem, based on partition theory, was introduced by Hartmanis[11], and is presented in detail in Hartmanis and Stearns[12] and Friedman and Menon[5]. The state minimization problem for completely specified machines was considered by Huffman[14], Moore[18] and Mealy[16]. The reduction problem for incompletely specified machines has been considered by Ginsburg[7,8], Paull and Unger[19] and Grasselli and Luccio[9,10]. The topic of sequential machine experiments was originally considered by Moore[18] and is thoroughly covered by Gill[6]. The subject of asynchronous sequential circuits is covered in great detail by Unger[21] and also Friedman and Menon[5]. The Turing machine is named after A. M. Turing[20] and extensive material on this subject is presented by Hopcroft and Ullman[13].

## REFERENCES

[1]  Armstrong, D. B., "A Programmed Algorithm for Assigning Internal Codes to Sequential Machines," *IRE Trans. Electronic Computers,* vol. EC-11, pp 466-472, August, 1962.

[2]  Armstrong, D. B., "On the Efficient Assignment of Internal Codes to Sequential Machines," *IRE Trans. Electronic Computers,* vol. EC-11, pp 611-622, October, 1962.

[3]  Cadden, W. J., "Equivalent Sequential Circuits," *IRE Transactions on Circuit Theory,* vol. CT-6, March 1959.

[4]  Dolotta, T. A., and E. J. McCluskey, "The Coding of Internal States of Sequential Circuits," *IEEE Trans. on Electronic Computers,* vol. EC-13, pp 549-562, October, 1964.

[5]  Friedman, A. D., and P. R. Menon, *Theory and Design of Switching Circuits,* Computer Science Press, Woodland Hills, Calif., 1975.

[6]  Gill, A., *Introduction to the Theory of Finite-State Machines,* McGraw-Hill, New York, 1962.

[7]  Ginsburg, S., "A Synthesis Technique for Minimal State Sequential Machines," *IRE Trans. Electron. Computers,* vol. EC-8, no. 1, pp. 13-24, March, 1959.

[8] Ginsburg, S., "On the Reduction of Superfluous States in a Sequential Machine," *J. Assoc. Computing Machinery*, vol. 6, pp. 259–282, April, 1959.

[9] Grasselli, A., and F. Luccio, "A method for minimizing the number of internal states in incompletely specified sequential networks," *IEEE Trans. on Electronic Computers*, vol. EC-14, pp 350–359, June, 1965.

[10] Grasselli, A., and F. Luccio, "A Method for Combined Row-Column Reduction of Flow Tables," *IEEE Conf. Record 1966 Seventh Symposium Switching and Automata Theory*, pp. 136–147, October, 1966.

[11] Hartmanis, J., "On the State Assignment Problem for Sequential Machines I," *IRE Trans. Electronic Computers*, vol. EC-10, pp 157–165, June, 1961.

[12] Hartmanis, J. and R. E. Stearns, *Algebraic Structure Theory of Sequential Machines*, Prentice-Hall Inc., Englewood Cliffs, N.J., 1966.

[13] Hopcroft, J. E. and J. D. Ullman, *Formal Languages and Their Relation to Automata*, Addison-Wesley, Reading, Mass., 1969.

[14] Huffman, D. A., "The Synthesis of Sequential Switching Circuits," *J. Franklin Institute.*, vol 257, pp 161–190, March, 1954, pp 275–303, April, 1954.

[15] McCluskey, E. J., and S. H. Unger, "A Note on the Number of Internal Variable Assignments for Sequential Switching Circuits," *IRE Trans. on Electronic Computers*, vol EC-8, pp 439–440, December, 1959.

[16] Mealy, G. H., "A Method for Synthesizing Sequential Circuits," *Bell System Technical Journal*, vol. 34, pp 1045–1079, September, 1955.

[17] Millman, J. and H. Taub, *Pulse, Digital, and Switching Waveforms*, McGraw-Hill, New York, N.Y., 1965.

[18] Moore, E. F., "Gedanken-experiments on Sequential Machines," pp 129-153, *Automata Studies*, Annals of Mathematical Studies, no. 34, Princeton University.

[19] Paull, M. C., and S. H. Unger, "Minimizing the Number of States in Incompletely Specified Sequential Switching Functions," *IRE Trans. Electron. Computers*, vol. EC-8, pp. 356-366, September, 1959.

[20] Turing, A. M. "On Computable Numbers," *Proceedings of London Mathematical Society*, vol 42, pp 230-265, 1936.

[21] Unger, S. H., *Asynchronous Sequential Switching Circuits*, John Wiley, New York, 1969.

## PROBLEMS

**5.1)** The circuit shown below is proposed for a new type of *FF*. Derive its transition table and evaluate its usefulness compared with *D*, *RS*, *JK FF*'s.

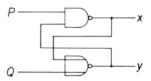

Figure 5.52   Problem 5.1

**5.2)** a) Derive a state table for the circuit of Figure 5.53. Assume all *FF*'s are clocked.

   b) Derive a circuit from the state table of (a) using the same state assignment but using *SR* flip-flops as memory elements.

**5.3)**   A sequential circuit for the state table of Figure 5.54 is to be constructed using *D* flip-flops as memory elements and one of the two state assignments shown. Which assignment is more economical in terms of hardware requirements? Assume two level sum of products logic.

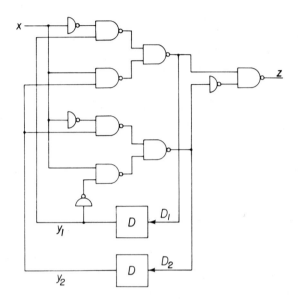

Figure 5.53   Problem 5.2

| | x | | State Assign. #1 | | State Assign. #2 | |
| | 0 | 1 | $y_1$ | $y_2$ | $y_1$ | $y_2$ |
|---|---|---|---|---|---|---|
| $q_1$ | $q_3,0$ | $q_2,1$ | 0 | 0 | 0 | 0 |
| $q_2$ | $q_4,1$ | $q_1,0$ | 0 | 1 | 0 | 1 |
| $q_3$ | $q_1,0$ | $q_4,0$ | 1 | 0 | 1 | 1 |
| $q_4$ | $q_2,1$ | $q_3,0$ | 1 | 1 | 1 | 0 |

Figure 5.54   Problem 5.3

**5.4)** a) Derive a state table for the circuit of Figure 5.55. Assume all *FF*'s are clocked.

b) Give a word description of the sequential function realized by this circuit.

**5.5)** Derive a state table for the circuit of Figure 5.56.

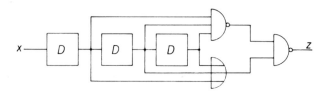

**Figure 5.55   Problem 5.4**

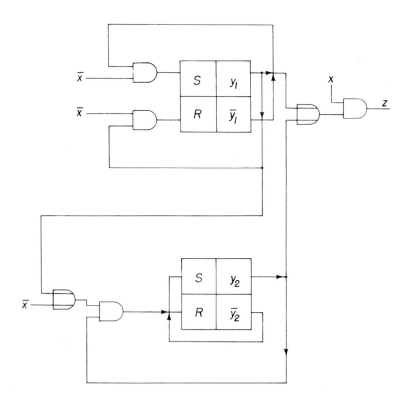

**Figure 5.56   Problem 5.5**

**5.6)** Reduce the following state table to a minimum state equivalent table.

|  | $x$ | |
|---|---|---|
|  | 0 | 1 |
| $q_1$ | $q_3,0$ | $q_2,1$ |
| $q_2$ | $q_3,1$ | $q_1,1$ |
| $q_3$ | $q_5,0$ | $q_2,1$ |
| $q_4$ | $q_6,0$ | $q_1,0$ |
| $q_5$ | $q_1,0$ | $q_7,1$ |
| $q_6$ | $q_4,1$ | $q_3,1$ |
| $q_7$ | $q_5,1$ | $q_3,1$ |

Figure 5.57   Problem 5.6

**5.7)** Derive a state table to operate as follows: there is one input $x$ and one output $z$. The initial state $(t = 0)$ is $q_0$. Let $U(t) =$ the number of 1-inputs to the circuit from $t = 0$ to $t$ and $V(t)$ = the number of 0-inputs from $t = 0$ to $t$. The output at time $t$, should be 1 if and only if $V(t) - U(t) = 3k$ for $k$ an integer (i.e., $V(t) - U(t) = ... -9, -6, -3, 0, 3, 6, 9, ....$ Reduce the table.

**5.8)** Using $T$ flip-flops as memory devices design a sequential circuit to realize the following state table, using the state assignment shown to the right of the table. The combinational logic realizing the $FF$ excitation should be minimal two level sum of products.

|  | $x$ | | | $y_1$ | $y_2$ |
|---|---|---|---|---|---|
|  | 0 | 1 | | | |
| $q_1$ | $q_3$ | $q_1$ | | 0 | 0 |
| $q_2$ | $q_2$ | $q_3$ | | 0 | 1 |
| $q_3$ | $q_4$ | $q_2$ | | 1 | 1 |
| $q_4$ | $q_2$ | $q_1$ | | 1 | 0 |

Figure 5.58   Problem 5.8

**5.9)** Consider the following sequential circuit. Obtain an equivalent circuit which requires less than half the number of gates.

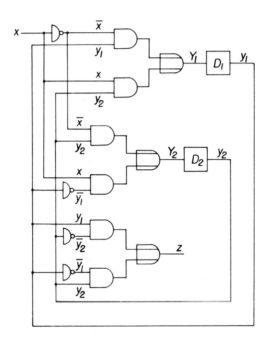

Figure 5.59   Problem 5.9

**5.10)** Derive a state table for a sequential circuit which has one input $x$, and one output $z$. The initial state is $q_0$ and the first two outputs should be 0. After that, the output at time $t$ is 1 if and only if the inputs at time $t$, $t - 1$, $t - 2$ contain exactly two 1-inputs and one 0-input. Reduce the table.

**5.11)** Reduce the state table of Figure 5.60 to an equivalent minimum state version.

**5.12)** a) Design a sequential circuit to operate as a constant multiplier as follows. The machine has one input $x$ and one ouptut $z$. If the input is interpreted as a serial representation of the number $x$ (low order bits first) the output sequence should represent

$$x$$

|       | 0        | 1        |
|-------|----------|----------|
| $q_1$ | $q_5,0$  | $q_2,1$  |
| $q_2$ | $q_6,0$  | $q_4,0$  |
| $q_3$ | $q_5,0$  | $q_2,1$  |
| $q_4$ | $q_6,0$  | $q_2,0$  |
| $q_5$ | $q_3,0$  | $q_6,1$  |
| $q_6$ | $q_2,0$  | $q_3,0$  |

**Figure 5.60   Problem 5.11**

the number $2 \cdot x$ (low order bits first).
b) Same as above but output represents $3 \cdot x$.

**5.13)** Construct the state diagram for a machine with one input, $x$, one output $z$, and $z = 1$ if the previous 4 inputs and the present input constitute a sequence of 5 symbols with exactly three 1's, two 0's and the first two symbols in the sequence are 11.

**5.14)** Design a sequential network using *JK* flip-flops with 2 control signal inputs $x_1, x_2$, which operates as specified below and counts modulo 8.

| $x_1 x_2$ | Operation |
|-----------|-----------|
| 0  0 | No change of state |
| 0  1 | Acts as counter, increments by 2 |
| 1  1 | Acts as counter, increments by 1 |
| 1  0 | Acts as counter, decrements by 1 |

**5.15)** For each of the following functions specify a finite state table to realize that function if possible to do so or prove it is impossible.
a) The output of the machine is 1 if and only if at least 2 of the previous 3 inputs were zero.
b) The output of the machine at time $t$ is 1 if and only if for the input sequence up to and including $t$, the number of odd length 1 sequences is even.

c) The output of the machine at time $t$ is 1 if and only if for the input sequence up to and including $t$, the length of the largest 0 sequence is even.

**5.16)** Use *JK* flip-flops as memory elements and design a modulo 8 counter which counts in *Gray Code* as shown below.

| Decimal Number | Gray Code Number | | |
|:---:|:---:|:---:|:---:|
| 0 | 0 | 0 | 0 |
| 1 | 0 | 0 | 1 |
| 2 | 0 | 1 | 1 |
| 3 | 0 | 1 | 0 |
| 4 | 1 | 1 | 0 |
| 5 | 1 | 1 | 1 |
| 6 | 1 | 0 | 1 |
| 7 | 1 | 0 | 0 |

**5.17)** Find a good state assignment for the following state table.

|  | $x$ | |
|:---:|:---:|:---:|
|  | 0 | 1 |
| $q_1$ | $q_2,0$ | $q_1,1$ |
| $q_2$ | $q_4,0$ | $q_3,0$ |
| $q_3$ | $q_2,0$ | $q_3,1$ |
| $q_4$ | $q_5,1$ | $q_5,1$ |
| $q_5$ | $q_1,0$ | $q_2,0$ |

Figure 5.61   Problem 5.17

**5.18)** (a) Specify procedures for deriving distinguishing sequences and homing sequences.
(b) Prove that every reduced machine has a homing sequence.
(c) Show that not all reduced machines have distinguishing sequences.

**5.19)** Modify Procedure 5.1 so that the outputs of the Mealy machine occur one time period after the corresponding Moore machine outputs.

# part 3

# Digital System Design and Related Problems

# CHAPTER 6:

# *System Level Design*

Up to now we have considered the design of *circuits* in which the basic elements, gates and *FF*'s were interconnected in such a manner as to realize functions which were specified by a state table (or truth table). In this chapter we will consider the design of digital systems in which the basic elements are related sets of *FF*'s called *registers*. Combinational and sequential circuits are used to perform computations on the contents of these registers under the control of other circuits. In system level design we are primarily interested in how these subsystems may be defined and interconnected so as to properly execute a sequence of information processing tasks rather than the detailed synthesis of the particular subsystems. Frequently the design of systems to perform relatively complex computations is simplified considerably if such design is done on the system level rather than the circuit level (as considered in the previous chapters). This is especially true for complex computations such as binary multiplication which can be done by a sequence of relatively simple information processing operations, register shifting and register addition. We will assume that operation is synchronous, all operations being controlled by a master clock so as to occur only at fixed clock times.

## 6.1 DESIGN LANGUAGE

In circuit level design, truth tables or state tables are used to specify the function to be designed. Languages which can be used to represent the flow of information on a system level are helpful in system level design. Such languages are frequently referred to as *Register Transfer Languages*. In this chapter we will use a language which may be considered to be a prototype of many similar languages. Our choice of this specific language is based solely on the relative ease of representing and interpreting the specification of the system designs we will be considering herein. The essential features of this language are as follows.

(1) *Representation of Systems Inputs, Outputs and Registers*
    (a) The inputs of a system will be denoted by $X$ (or $Y, V, W$) and a particular input bit by $x_i$ (or $y_i, v_i, w_i$). A constant input will be represented by a number $n$.
    (b) The outputs of a system will be denoted by $Z$, and a particular output by $z_i$.
    (c) The registers of a system will be denoted by capital letters (excluding those letters used to represent inputs or outputs) or by a sequence of capital letters such as $A$, $MBR$, etc. A particular *FF* (bit) of register $R$ will be denoted by $R_i$. We assume that the bits of an $n$-bit register are numbered from right to left as shown in Figure 6.1. A set of consecutive bits $R_i, R_{i+1}, ..., R_j$ will be denoted by $R_{i,j}$.

**Figure 6.1   A register**

(2) *Basic System Operations*
    The basic system operations result in a modification of the contents of some register or registers. Some commonly used operations are as follows.
    (a) Setting register $R$ to the value of an input $X$:

$$R \leftarrow X$$

It is assumed that if $R$ has bits $R_0, ..., R_{n-1}$ then $X$ consists of $x_0, ..., x_{n-1}$ and bit $R_i$ of $R$ is set to $x_i$.
    (b) Setting register $R$ to the value of a constant number $k$:

$$R \leftarrow k$$

This results in $R$ assuming the binary representation of $k$.
(c) Transferring the contents of register $B$ to register $A$:

$$A \leftarrow B.$$

It is assumed that $A$ and $B$ have the same number of bits and $B_i$ is transferred to $A_i$ for all $i$.
(d) Perform an arithmetic operation on the contents of two registers and store the result in another register (or one of the operand registers). Some commonly used operations are:
   (i) Addition: $C \leftarrow A + B$. ($A$ and $B$ are added bit by bit and the result is stored in $C$.)
   (ii) Subtraction: $C \leftarrow A - B$.
   (iii) Multiplication: $C \leftarrow A \cdot B$.
   (iv) Division: $C \leftarrow A/B$.
(e) Perform a logical operation on the contents of a register or registers and store the result in another register (or one of the operand registers). Some commonly used operations are:
   (i) Logical OR: Since the symbol $(+)$ for logical OR has been used for arithmetic addition we will use the symbol $\vee$.

$$C \leftarrow A \vee B$$

   (The logical OR of bits $A_i$ and $B_i$ is stored in $C_i$ for all $i$.)
   (ii) Logical AND: Since the symbol $(\cdot)$ for logical AND has been used for arithmetic product we will use the symbol $\wedge$.

$$C \leftarrow A \wedge B$$

   (The logical AND of bits $A_i$ and $B_i$ is stored in $C_i$ for all $i$.)
   (iii) Complementation: $B \leftarrow \bar{A}$. (The complement of bit $A_i$ is stored in $B_i$ for all $i$.)
(f) Perform an operation which shifts the contents of a register.
   (i) Right shift: $A \leftarrow SR(A)$; results in $A_i \leftarrow A_{i+1}$ for $0 \le i \le n - 2$, $A_{n-1} \leftarrow 0$.
   (ii) Left Shift: $A \leftarrow SL(A)$; results in $A_i \leftarrow A_{i-1}$ for $1 \le i \le n - 1$, $A_0 \leftarrow 0$.

(iii) Right Circulate: $A \leftarrow RCIRC(A)$; results in $A_i \leftarrow A_{i+1}$
for $0 \leq i \leq n - 2$, $A_{n-1} \leftarrow A_0$.

(iv) Left Circulate: $A \leftarrow LCIRC(A)$; results in $A_i \leftarrow A_{i-1}$
for $1 \leq i \leq n - 1$, $A_0 \leftarrow A_{n-1}$.

## Example 6.1

Let $A$ and $B$ be 4-bit registers whose contents are 0110 and 0101 respectively.

The operation $C \leftarrow A \vee B$ results in the contents of $C$ of 0111. The operation $D \leftarrow A \wedge B$ results in the contents of $D$ becoming 0100. The operation $E \leftarrow \bar{A}$ results in the contents of $E$ becoming 1001.

### (3) Timing Considerations

The system is assumed to operate in an essentially synchronous manner, controlled by clock pulses occurring at times $T_i$, $i = 1, 2, \ldots, p$. Associated with each operation is a specific time at which it occurs which is denoted as follows:

$T_i$: Operation.

In this way the order in which operations are executed can be specified. An operation which occurs at $T_i$ precedes one at $T_j$ if $i < j$. Several operations may be specified to occur at the same clock pulse as follows:

$T_i$: Operation #1; Operation #2; ... etc.

Our language has now been specified to a sufficient extent so that many algorithms may be specified as a sequence of fundamental register operations.

## Example 6.2

(a) The algorithm to compute the 2's complement of input $X$ using Procedure 2.5 is specified as follows:

$$T_1: \quad A \leftarrow X; B \leftarrow 1$$

$$T_2: \quad A \leftarrow \bar{A}$$

$$T_3: \quad C \leftarrow A + B$$

(b) Compute $X + Y - V/W$.

$$T_1: \quad A \leftarrow X; B \leftarrow Y; C \leftarrow V; D \leftarrow W$$

$$T_2: \quad A \leftarrow A + B; C \leftarrow C/D$$

$$T_3: \quad A \leftarrow A - C$$

## 6.2 SYSTEM AND CONTROL UNIT DESIGN

### 6.2.1 SIMPLE CONTROL CIRCUITS

In order to have a system perform such a sequence of operations, it is necessary to distinguish between different occurrences of the clock pulse. A special circuit, called a *sequencer* or *control circuit* is used to perform this task. This unit is a sequential circuit, which controls the order in which a sequence of operations are executed. To control a sequence of operations occurring on $n$ consecutive clock pulses, the control unit effectively counts the first $n$ clock pulses and translates them into distinct output control signals $S_1, S_2, ... S_n, S_i$ corresponding to clock pulse $T_i$. These signals are transmitted to the subnetworks which perform the operations and are used to control the sequence in which these operations are executed. After completing this task the control circuit returns to its initial state and can then control the same operation sequence on different data. A control circuit which always transmits the same sequence of control signals in this manner is called a *simple control circuit*. A block diagram of a digital system which functions in this manner is shown in Figure 6.2.

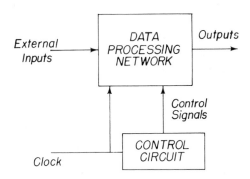

Figure 6.2   System with simple control circuit

A simple control circuit is in effect a one column sequential machine, whose only input is the master clock. The machine in effect counts clock pulses modulo $n$ and outputs a control signal $S_i$ for clock pulse $T_k$, if $k = i \bmod n$. Such a circuit is a modulo $n$ counter whose state table is shown in Figure 6.3.

Clock

$$
\begin{array}{c|l}
q_0 & q_1 , S_1 \\
q_1 & q_2 , S_2 \\
q_2 & q_3 , S_3 \\
\cdot & \\
\cdot & \\
\cdot & \\
\cdot & \\
q_{n-2} & q_{n-1} , S_{n-1} \\
q_{n-1} & q_0 , S_0
\end{array}
$$

Figure 6.3   Simple control unit state table

The outputs of the control unit, the *control signals*, must be generated as a set of binary signals, $\lceil \log_2 n \rceil$ such signals being required to generate $n$ distinct control signals. If these binary signals are input to a decoder (such as those considered in Section 4.1), a single output signal $S_i$ is generated for each distinct control signal (Figure 6.4). Once the control unit for an algorithm has been specified, the data processing network can be designed. This network uses the appropriate control signals as gating or enabling signals, as illustrated in Section 5.7, to control the execution of the individual data processing operations.

Thus a digital system to perform an algorithm may be designed as outlined in the following procedure.

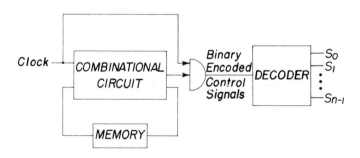

Figure 6.4   Representation of simple control circuit

**Procedure 6.1**

(1) Specify the algorithm as a sequence of *basic* operations in a Register Transfer Language. (The use of a basic operation, implies the existence of a subsystem which can perform that operation in a single clock period).

(2) Specify the (simple) control unit to generate the correct control signal sequence to enable correct execution of the algorithm.

(3) Design the data processing network using the appropriate control signals as enabling signals to control the execution of the individual data processing operations.

**Example 6.3**

We will design an *input-output buffer* (IOB) to operate as follows. Data is transmitted to the IOB in 8-bit groups called bytes. The IOB accumulates four successive bytes and then transmits them as a 32-bit word to the computer. The general system for the IOB is shown in Figure 6.5.

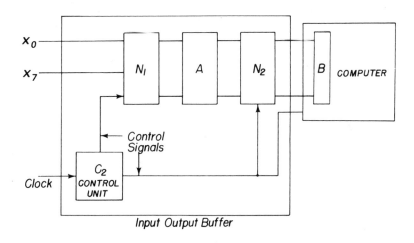

Figure 6.5   Input-output buffer

The 8 bit bytes occur on inputs $x_0, x_1, ..., x_7$, and are accumulated and stored in $A$, a 32-bit register of $JK$ flip-flops, under the control of $C_2$. Then a control signal is generated which gates the 32-bit word to the $B$ register of the computer. The circuit $N_1$ is a combinational circuit which uses the control signals to gate the inputs into the appropriate bit of $A$, and $N_2$ is a similar circuit which gates the 32-bit word to the computer under the control of $C_2$.

The following sequence of basic operations must be executed.

$$T_1: \quad A_{0,7} \leftarrow X$$
$$T_2: \quad A_{8,15} \leftarrow X$$
$$T_3: \quad A_{16,23} \leftarrow X$$
$$T_4: \quad A_{24,31} \leftarrow X$$
$$T_5: \quad B \leftarrow A$$

The control unit must generate a sequence of 5 control signals to control the execution of these operations. Such a control unit is defined by the state table of Figure 6.6.

Clock

| | |
|---|---|
| $q_0$ | $q_1,S_1$ |
| $q_1$ | $q_2,S_2$ |
| $q_2$ | $q_3,S_3$ |
| $q_3$ | $q_4,S_4$ |
| $q_4$ | $q_0,S_0$ |

Figure 6.6   IOB control unit state table

The network $N_1$ is a combinational circuit which generates the excitations for the $FF$'s of register $A$. These excitation functions are of the form

$$\left. \begin{array}{l} J_i = S_1 \cdot x_i \\ K_i = S_1 \cdot \bar{x}_i \end{array} \right\} \quad 0 \le i \le 7$$

$$\left. \begin{array}{l} J_i = S_2 \cdot x_{(i-8)} \\ K_i = S_2 \cdot \bar{x}_{(i-8)} \end{array} \right\} \quad 8 \le i \le 15$$

$$\left. \begin{array}{l} J_i = S_3 \cdot x_{(i-16)} \\ K_i = S_3 \cdot \bar{x}_{(i-16)} \end{array} \right\} \quad 16 \le i \le 23$$

$$J_i = S_4 \cdot x_{(i-24)}$$
$$K_i = S_4 \cdot \bar{x}_{(i-24)}$$
$$\left.\right\} \quad 24 \leq i \leq 31$$

The network $N_2$ is a combinational circuit which generates the excitations for the $B$ register. These are of the form

$$J_i = S_0 \cdot y_i$$
$$K_i = S_0 \cdot \bar{y}_i$$
$$\left.\right\} \quad 0 \leq i \leq 31$$

where $y_i$ is the state of $A_i$.

**Example 6.4**

Another computation which can be performed under the control of a simple control circuit is the 2's complement. We shall now design a digital system which computes the 2's complement of input $X$. An algorithm for this computation can be expressed in our register transfer language as:

$$T_1: \quad A \leftarrow X; B \leftarrow 1$$
$$T_2: \quad A \leftarrow \bar{A}$$
$$T_3: \quad A \leftarrow A + B$$

The execution of this algorithm requires a control unit to generate a sequence of thrèe control signals repetitively. Such a control unit is described by the state table of Figure 6.7. The use of the addition operation

Clock

| | |
|---|---|
| $q_0$ | $q_1, S_1$ |
| $q_1$ | $q_2, S_2$ |
| $q_2$ | $q_0, S_0$ |

Figure 6.7   Control unit state table for 2's complement addition

in the algorithm implies the existence of a subsystem which can perform this operation within a single clock period. The entire system can be represented schematically as shown in Figure 6.8.

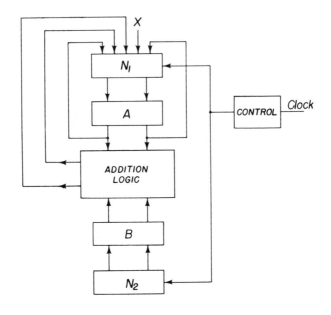

**Figure 6.8    System for 2's complement addition**

The excitation inputs to the $A$ register are generated by $N_1$ and are such as to generate the state behavior specified by:

$$Y_{A_i} = S_1 \cdot x_i + S_2 \bar{y}_{A_i} + S_0 \cdot z_i, 1 \leq i \leq n$$

where $z_i$ are the outputs of the addition logic. Similarly, $N_2$ generates the excitation inputs to the $B$ register in accordance with the state behavior

$$Y_{B_i} = 0, i > 1$$

$$Y_{B_1} = S_1.$$

### 6.2.2 STATUS DEPENDENT CONTROL UNITS

A simple control unit generates a repetitive sequence of control signals and hence can be used for repetitive execution of certain types of operations. However, some types of computations may require a sequence of control signals which may depend upon and vary with the specific data upon which the computation is performed. In this case the control unit must receive input information from the data processing network to generate the control sequences. Such a controller is said to be a *status dependent controller*. A general system with a status dependent controller is as shown in Figure 6.9.

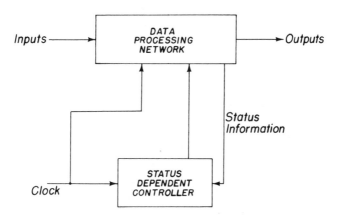

Figure 6.9  System with status dependent control unit

As an example of an information processing task which requires a status dependent controller consider the design of a digital system to compute the square root of $N$. There exists an algorithm which computes the square root of a number through a process of convergence. This algorithm uses the arithmetic operations of addition, division, and comparison. The algorithm, which computes the square root to within an accuracy specified by a constant, $k$, is as follows.

(1) Approximate the square root of $N$ as $N/2 = X_o$ as a first approximation.

(2) Take a next approximation as $X_1 = 1/2 \, (X_o + N/X_o)$.

(3) Continue calculating the $(i + 1)^{st}$ approximation as

$$X_{i+1} = \frac{1}{2}\left(X_i + \frac{N}{X_i}\right)$$

where $X_i$ is the $i^{th}$ approximation. The algorithm terminates when two successive approximations differ by $< k$, a constant.

If $N = 6$ and $k = .2$, the computation of the square root of $N$ requires two iterations of this procedure and is as follows:

1st Approx:     $X_o = 3$

2nd Approx:     $X_1 = \dfrac{1}{2}\left(3 + \dfrac{6}{3}\right) = 2.5$

3rd Approx:     $X_2 = \dfrac{1}{2}\left(2.5 + \dfrac{6}{2.5}\right) = \dfrac{1}{2}(4.9) = 2.45$

However, if $N = 20$ and $k = .2$, three iterations are required.

1st Approx:     $X_o = 10$

2nd Approx:     $X_1 = \dfrac{1}{2}\left(10 + \dfrac{20}{10}\right) = 6$

3rd Approx:     $X_2 = \dfrac{1}{2}\left(6 + \dfrac{20}{6}\right) = \dfrac{1}{2}(9.33) = 4.67$

4th Approx:     $X_3 = \dfrac{1}{2}\left(4.67 + \dfrac{20}{4.67}\right) = \dfrac{1}{2}(8.96) = 4.48$

In order to enable us to specify such an algorithm in our register transfer language we must add new expressions to the language. Associated with each operation is a clock pulse $T_i$ and an associated control signal $S_i$. By adding the expression GO TO $T_i$ (or GO TO $S_i$), we can repeat a sequence of operations an indefinite number of times while only explicitly writing them once. In order to be able to use status information to determine which operation to perform we add an expression of the form

IF (STATUS CONDITION) THEN CONTROL SIGNAL 1:

$\left.\begin{array}{l} \text{OPERATION} \\ \text{EXPRESSION} \end{array}\right\}$ 1, ELSE CONTROL SIGNAL 2:

$\left.\begin{array}{l} \text{OPERATION} \\ \text{EXPRESSION} \end{array}\right\}$ 2 which causes operation 1 to be executed if the status condition is satisfied while operation 2 is executed if the status condition is not satisfied.

We can now specify the procedure to compute the square root of $X$ in our register transfer language as follows:

|  |  | Explanation |
|---|---|---|
| $S_1$: | $A \leftarrow X, C \leftarrow k$ | Store input $x$ in Register $A$ and constant $k$ in Register $C$ |
| $S_2$: | $AP \leftarrow SR[A]$ | $X/2$ is input to $AP$(1ˢᵗ approximation) (Note that division by 2 is equivalent to a right shift) |

$$\left. \begin{array}{ll} S_3: & B \leftarrow A/AP \\ S_4: & B \leftarrow AP + B \\ S_5: & B \leftarrow SR(B) \end{array} \right\}$$   generate next approximation

$S_6$:  IF $(B \geq AP)$ THEN $S_7$:      Compute difference with previous approximation and store in $D$.

$\quad D \leftarrow B - AP$ ELSE $S_8$:

$\quad D \leftarrow AP - B$

$S_9$:  IF $(D < C)$ THEN $S_0$:      If $D < k$, output result, and prepare for next computation. If $D \geq k$, begin computation of next approximation.

$\quad OUTPUT \leftarrow B$;   GO TO $S_1$,

$\quad$ ELSE $S_{10}$:   $AP \leftarrow B$;

$\quad$ GO TO $S_3$

This procedure must be executed under the control of a status dependent control unit. Such a control unit can be specified as a finite state sequential machine whose input columns correspond to the possible values of the status inputs which affect the generation of control signals. In the square root algorithm there are two status dependent operations both of which are the result of the comparison of the contents of two registers. The use of these instructions requires the existence of a comparison subcircuit. We assume that this circuit generates an output $z = 1$ if $B \geq AP$ and $z = 0$ if $B < AP$ and similarly $z = 1$ if $D \geq C$ and $z = 0$ if $D < C$. This output $z$ is then the status input of the control unit. The sequential machine of Figure 6.10 will generate the appropriate sequence of control signals (assuming initial state $q_0$) for the execution of the square root algorithm previously specified in register transfer language form. Note

$$z$$

| | 0 | 1 |
|---|---|---|
| $q_0$ | $q_1, S_1$ | $q_1, S_1$ |
| $q_1$ | $q_2, S_2$ | $q_2, S_2$ |
| $q_2$ | $q_3, S_3$ | $q_3, S_3$ |
| $q_3$ | $q_4, S_4$ | $q_4, S_4$ |
| $q_4$ | $q_5, S_5$ | $q_5, S_5$ |
| $q_5$ | $q_6, S_6$ | $q_6, S_6$ |
| $q_6$ | $q_7, S_8$ | $q_7, S_7$ |
| $q_7$ | $q_8, S_9$ | $q_8, S_9$ |
| $q_8$ | $q_0, S_0$ | $q_2, S_{10}$ |

**Figure 6.10   Control unit for square root system**

that the next state and output for states $q_0$ through $q_4$, which correspond to the first 5 instructions in the algorithm are the same for $z = 0$ and $z = 1$ and hence are independent of the status $z$. In state $q_5$ the control signal $S_6$ is generated and this causes $B$ and $AP$ to be compared. Depending on the result of this comparison, if $B \geq AP$, $z = 1$, and control signal $S_7$ is generated in state $q_6$, or if $B < AP$, $z = 0$ and $S_8$ is generated. Then in state $q_7$, control signal $S_9$ is generated which causes $D$ and $C$ to be compared. If $D \geq C$, $z = 1$, and control signal $S_{10}$ is generated and the next iteration started by the transition $N(q_8, 1) = q_2$. If $D < C$, $z = 0$ and control signal $S_0$ is generated and the next computation initiated by the transition $N(q_8, 0) = q_0$. The general structure of the system is shown in Figure 6.11.

## 6.2.3 INSTRUCTION DEPENDENT CONTROL UNITS

So far we have considered the design of control networks which can carry out a single data processing task. For each data processing task requiring a different control signal sequence generation a new control unit must be designed. However, it is possible to design a single control network which can generate control signal sequences for several information processing operations. Such a controller is referred to as an *instruction-dependent control unit*. This controller must have inputs that specify which instruction is currently being executed. In response to

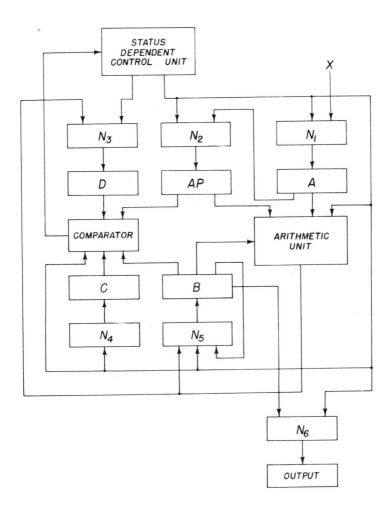

**Figure 6.11   Square root system**

the inputs specifying the instruction, and the status inputs (if any) the controller generates the appropriate control signal sequence. A general system with an instruction dependent control unit is shown in Figure 6.12. The control signals may be input to the instruction source to ensure that the instruction information changes at the appropriate time. Specifying the control unit and reducing the number of states is essentially

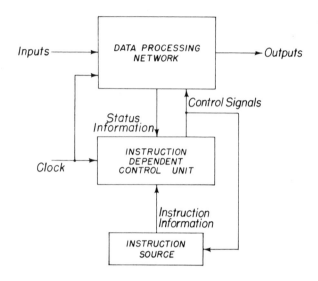

Figure 6.12   System with instruction dependent control unit

the same as deriving and reducing the state table for any sequential machine.* However, minimizing the number of outputs in the control unit is an added problem, as illustrated in the following example.

## Example 6.5

We will design a control unit to generate the appropriate control sequence to execute either of two instructions as specified by the value of an input $I$. If $I = 0$ the two's complement of $X$ is computed. If $I = 1$, the complement of $X + 1$ will be computed. We can express an algorithm for both of these instructions in our *RTL* as follows:

$$I = 0$$
$$S_1: \quad A \leftarrow X; B \leftarrow 1$$
$$S_2: \quad A \leftarrow \bar{A}$$

*The instruction dependent control units we will consider are incompletely specified sequential machines whose unspecified entries correspond to prohibited input sequences such as in the table of Figure 5.3. For such machines state reduction procedures which are similar to those for completely specified sequential machines can be utilized to minimize the number of states in the control unit.

$$S_0: \quad C \leftarrow A + B$$

$$I = 1$$

$$S_3: \quad A \leftarrow X; B \leftarrow 1$$

$$S_4: \quad C \leftarrow A + B$$

$$S_5: \quad C \leftarrow \bar{C}$$

The state table for this controller is shown in Figure 6.13(a), where we have assumed that $I$ can only change value when the control unit is in state $q_0$.

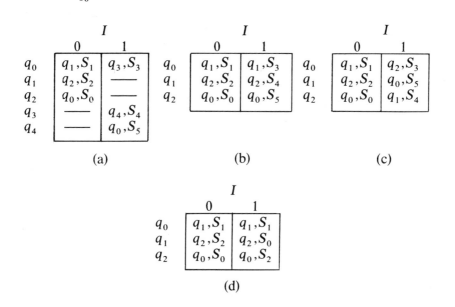

Figure 6.13    Control unit for Example 6.5

If $I = 0$, the control unit passes through states $q_1$ and $q_2$ before returning to initial state $q_0$ and generates the control signal sequence $S_1 S_2 S_0$. If $I = 1$, the control unit passes through states $q_3$ and $q_4$ before returning to $q_0$ and generates the control signal sequence $S_3 S_4 S_5$. We assume that the input $I$ only changes value when the control unit is in state $q_0$ so the entries in states $q_1$ and $q_2$ in column 1 and $q_3, q_4$ in column 0 are left unspecified. In effect, in state $q_0$ the only allowed input sequences are 000 or 111. This table can be reduced by combining states $q_1, q_3$ and $q_2, q_4$ to that of Figure 6.13(b) or by combining states $q_1, q_4$ and $q_2, q_3$ to that of Figure 6.13(c). It is also possible to reduce the number of distinct control signals which must be generated. Note that $S_1$ and $S_3$ produce the same logic operations $A \leftarrow X$ and $B \leftarrow 1$. Therefore, $S_3$ can be replaced by $S_1$.

Similarly, $S_0$ and $S_4$ produce the operation $C \leftarrow A + B$, so $S_4$ can be replaced by $S_0$. The signals $S_2$ and $S_5$ produce the operations $A \leftarrow \bar{A}$ and $C \leftarrow \bar{C}$ respectively. These operations affect different registers and hence may be considered to be independent. If we replace $S_5$ by $S_2$ the control signal $S_2$ will cause both $A \leftarrow \bar{A}$ and $C \leftarrow \bar{C}$ and the effective computation will not be changed. Applying these control signal reductions to the table of Figure 6.13(b) results in the reduced table of Figure 6.13(d).

## 6.3 DIGITAL COMPUTER MODEL

We have assumed that the inputs which specify the instruction to be executed are controlled by a source external to the control unit. However, this source must be synchronized with the control unit since these inputs are assumed to remain constant until the complete control signal sequence for the current instruction has been generated. Let us now consider an extension of this model in which the control unit generates a control signal which also controls the inputs which specify the instruction. (Recall Example 5.4, in which a sequential circuit which partially controls its own inputs has been considered). This extended capability of the instruction dependent control unit enables it to automatically sequence through a series of operations (called a *program*). These operations are assumed to be contained in a list which indicates the order in which these instructions must be executed. Such a system might be referred to as a *programmed computer*. In most computers the program is stored in a subunit of the computer called the memory unit. Thus the term *stored program computer*. The basic organization of such a system is as shown in Figure 6.14. The *Memory Unit* is used to store the program to be executed and the data upon which the program operates. The *Central Processing Unit (CPU)* consists of the registers and logic required to perform the basic logical and arithmetic operations which constitute a program. The *Control Unit* controls the sequence in which instructions are executed and also controls the CPU and other system

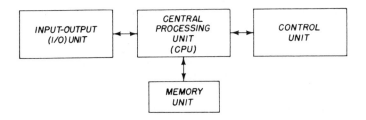

Figure 6.14    Block diagram of stored program computer

units during the execution of each instruction. The *I/O Unit* serves as an interface between the stored program processor and the outside world. We shall now consider the manner in which a program that is stored in the memory unit is executed.

Each instruction is stored in one word of memory. To execute the program the control unit must read an instruction from memory, determine the operation specified by the instruction, generate the sequence of control signals required to perform this operation, and then determine the location of the next instruction in memory, read that instruction, etc.

We shall first consider the process by which information in memory is located and read.

The memory unit is made up of $2^k$ $r$-bit registers. The contents of each register is called a *word*, and each register has a unique *address* associated with it to distinguish it from all the other registers. In most computer systems the memory registers are not designed from *FF's* but from *magnetic cores.** This is because large memories are required and magnetic cores are considerably cheaper than *FF's* (which are usually made of semiconductor material). If any register in the memory can be accessed at any time, the memory is referred to as a *random access memory.* We shall assume that the process of reading a register in a magnetic core memory results in the destruction of the contents of that

---

*A magnetic core is actually a small donut shaped piece of magnetic material which is capable of assuming one of two magnetized states depending on the direction of current excitation applied to the core via wires through its center. They are described in more detail elsewhere [2,3]. The important properties for our consideration are our assumptions that reading a memory register destroys the contents of that register and a register must be initialized to 0 (cleared) before writing in it. For some types of core these assumptions may not be valid.

register. Hence when reading a register in the memory unit, the control unit must restore the contents of that register after the destructive read. Before writing in a register, that register must be cleared (reset).

To accomplish this systematically, two specially designated registers called the *Memory Buffer Register (MBR)* and the *Memory Address Register (MAR)* are utilized. The *MAR* is used whenever memory is accessed (read or written into) to store the address (name) of the specific register in memory which is being accessed. The contents of this register, in the case of a read, or the contents to be written in this register, in the case of a write, are stored in the *MBR* which serves in effect as an interface between the *CPU* and the memory. In order to read a memory register $M_L$ located at address $L$, the following 3 step process is executed:

|   | Operation | Explanation |
|---|-----------|-------------|
| (1) | $MAR \leftarrow L$ | The address of the word to be accessed is stored in $MAR$ |
| (2) | $MBR \leftarrow M_{[MAR]};$ <br> $M_{[MAR]} \leftarrow 0$ | The contents of the memory register at the address specified by $MAR$ (which is denoted by $M_{[MAR]}$) is transferred to $MBR$ and that location is set to 0. |
| (3) | $M_{[MAR]} \leftarrow MBR$ | The contents of the $MBR$ is written into the memory register at address $L$. |

After completion of this process the desired word has been restored in its original memory location and is also available in the *MBR*. The process by which data in a register $D$ is written into the memory unit at location $L$ is similar to the reading process and consists of the following steps.

|   | Operation | Explanation |
|---|-----------|-------------|
| (1) | $MAR \leftarrow L;$ <br> $MBR \leftarrow D$ | The location to be written into is stored in the $MAR$ and the data to be written is stored in the $MBR$. |
| (2) | $M_{[MAR]} \leftarrow 0$ | The memory register is cleared. |
| (3) | $M_{[MAR]} \leftarrow MBR$ | The contents of the $MBR$ is written in the memory register $M_L$. |

After this process the specified address $L$ of the memory unit contains the data $D$.

The control unit has two basic phases of operation, the *fetch phase*, in which instructions are fetched from memory in the proper sequence, and the *execute phase* in which the instructions are executed by the control unit's generation of the proper sequence of control signals.

Each instruction word in memory is partitioned into disjoint groups of bits (called *fields*), which contain specific types of information. One such field, the operation code field, identifies the type of operation being performed. Assuming that the number of bits in the operation code field is constant (fixed field length) then $c$ bits in this field are adequate if the system has at most $2^c$ types of instructions. The remaining fields of an instruction word supply additional information required to execute the instruction. In the most general case this information includes the addresses of the two operands, the address where the result is to be stored, and the address where the next instruction to be executed is stored. Thus four address fields may be required. However, several possible simplifications can be used to reduce this requirement.

During the fetch phase, the control unit must keep track of the next instruction address. However, this address need not be specified in the instruction being executed, if the instructions of a program are stored in sequential address positions of the memory unit. In this case if the address of the current instruction is $I$, the address of the next instruction is $I + 1$, unless the current instruction is a *transfer* instruction, which specifies some other next instruction address. A special register, the *instruction address register (IAR)* is used to store the address of the next instruction to be executed. After each instruction is executed, the contents of the $IAR$ is incremented by one unless the instruction executed is a transfer operation. In this case the operand address fields can be used to specify the contents of the address of the next instruction, which is then stored in the $IAR$.

Thus the fetch phase consists of the following status dependent control sequence.

| Control Sequence | Explanation |
|---|---|
| $S_1: MAR \leftarrow IAR$ | The next instruction to be exe- |
| $S_2: MBR \leftarrow M_{[MAR]}$ | cuted, the location of which is |
| $S_3: M_{[MAR]} \leftarrow MBR$ | specified in the $IAR$, is fetched |
| $S_4: IR \leftarrow MBR_{i,j}$ | and the operation code bits |

| Control Sequence | Explanation |
|---|---|
| | $(MBR_{i,j})$ are stored in $IR$, the *Instruction Register*. |
| $S_5$: IF (IR indicates Transfer Operation) THEN | The address of the next instruction to be executed is stored in the $IAR$. This address is deter- |
| $S_6$: $IAR \leftarrow MBR_{ADD\ FIELD}$ ELSE | mined by incrementing the con- tents of $IAR$ by 1 unless the |
| $S_7$: $IAR \leftarrow IAR + 1$ | present instruction is a transfer operation in which case the next instruction is in the memory lo- cation whose address is specified in the current instruction address field in $MBR$. |

Thus the state table of the control unit for the fetch phase is as shown in Figure 6.15, where the output of a decoder is $T = 0$ if the present operation is not a transfer operation and $T = 1$ if it is a transfer operation.

During the execution phase the contents of the $IR$ and the values associated with related status signals determine the control signal sequence

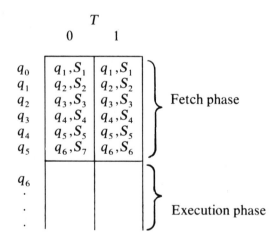

Figure 6.15   State table describing control unit fetch phase

generated, and then the control unit returns to the initial state of the fetch phase, $q_0$.

The result address field can be eliminated if the result of a particular type of operation is always stored in a special register (the particular register may depend on the operation). Similarly, one of the operand fields can be eliminated if for each type of operation one of the operands is always stored in a special register (which again may vary with the operation). Thus a single address field is adequate if all of these simplifying assumptions are used. The address field must consist of $k$ bits in order to specify $2^k$ possible memory locations. Alternatively, we can partition the memory unit into $2^{k'}$ segments of $2^{k-k'}$ words and specify a memory location by using a $k'$ bit register (frequently called an *index register*) to specify the segment in which the specific location is, and a $k-k'$ bit address field to specify the particular location within that segment. This reduces the number of bits required for the address field. However, additional instructions are required to transfer data to the index registers.

## 6.4 MACROINSTRUCTIONS AND MICROINSTRUCTIONS

Once the basic RTL instructions which must be performed, and the control signals which enable these operations have been specified, the CPU logic can be implemented using standard logic design techniques. The execution of a single computer instruction may require a complex sequence of RTL instructions as illustrated previously for the square root function. Such an instruction whose execution may require a sequence of control signals, is called a *macroinstruction*.

The basic RTL instructions which are assumed to be executable in one clock period and require one control signal are called *microinstructions*. The following sequence of steps define the fundamental design process of a programmed computer.

(1) Define basic macroinstructions.
(2) Define the sequence of logical and arithmetic microinstructions and the corresponding control signal sequence for each macroinstruction.
(3) Design control unit to generate the appropriate control signal sequence for each macroinstruction including access of next instruction.
(4) Design CPU to perform logical and arithmetic microinstructions

using appropriate control signals as enabling signals.

An important aspect of this procedure is the selection of an algorithm of microinstructions for each macroinstruction. The following example illustrates the specification of a sequence of RTL microinstructions for the arithmetic macroinstructions, multiplication and division.

### Example 6.6

Consider the design of a binary multiplier. Assume that we wish to form the product $X \cdot Y$ where $Y = y_n y_{n-1} \ldots y_o, y_i = 0,1$. Then

$$
\begin{aligned}
X \cdot Y &= X \cdot (y_n 2^n + y_{n-1} 2^{n-1} + \ldots + y_1 2^1 + y_o 2^o) \\
&= (X \cdot y_o) + 2(X \cdot y_1) + 2^2(X \cdot y_2) \ldots + 2^n(X \cdot y_n) \\
&= \sum_{i=0}^{n} 2^i (X \cdot y_i).
\end{aligned}
$$

However, since $y_i = 0,1$, $X \cdot y_i = X$ if $y_i = 1$, and $X \cdot y_i = 0$ if $y_i = 0$. Thus $2^i(X \cdot y_i) = 2^i X$ or 0. Furthermore, $2^i X$ is simply $X$ shifted to the left $i$ positions. Thus multiplication of two $n$-bit numbers can be performed by a sequence of $n$ shift and add operations. Such an algorithm can be specified as follows.

| Multiplication Algorithm | Explanation |
|---|---|
| $T_1$: $S_1$:   $A \leftarrow 0$; $M \leftarrow X$; $Q \leftarrow Y$; $D \leftarrow n$ | The two operands $X$ and $Y$ are stored in Registers $M$ and $Q$ respectively. The result register $A$ (which has $2n$ bits) is initialized to 0. The number of bits, $n$, is stored in $D$. |
| $T_2$:   IF $(Q_o = 1)$ THEN $S_2$: $A_{[n,2n-1]} \leftarrow A_{[n,2n-1]} + M$; ELSE GO TO $T_3$ | If the least significant bit of $Q$, $Q_o$, is 1, add $M$ to the $n$ most significant bits of $A$. Shift $A$ to the right one bit,* shift $Q$ to the right one bit |
| $T_3$:   $S_3$; $Q \leftarrow SR(Q)$; $A \leftarrow SR(A)$; $D \leftarrow D - 1$ | and decrement $D$ (independent of $Q_o$) |

---

*Alternatively, instead of shifting $A$ to the right after each iteration we could shift $M$ to the left. In this case $M$ must have $2n$ bits, and if $Q_i = 1$, $A \leftarrow A + M$.

| Multiplication Algorithm | Explanation |
|---|---|
| $T_4$: IF $(D = 0)$ THEN $S_0$: read next instruction; ELSE GO TO $T_2$. | This is iterated $n$ times until $D = 0$. |

The control unit state table for this algorithm is shown in Figure 6.16. In this algorithm we have used a microinstruction which decrements the contents of the $D$ register by one. Hence the counting part of the procedure is carried out by CPU logic. Alternatively the control unit could count up to $n$ and the $D$ register would not be needed. For instance, for $n = 3$ the state table of Figure 6.17 defines a control unit which generates the necessary signal sequence for a 3-bit multiplication.

The control unit in effect counts up to 3 by the state sequence $q_1, q_2, q_3$ and then generates the completion signal $S_0$ and returns to the initial state $q_0$.

|  | $D = 0$ | $D \neq 0$ $Q_o = 1$ | $D \neq 0$ $Q_o = 0$ |
|---|---|---|---|
| $q_0$ | $q_1, S_1$ | $q_1, S_1$ | $q_1, S_1$ |
| $q_1$ | $q_0, S_0$ | $q_2, S_2$ | $q_1, S_3$ |
| $q_2$ | — | $q_1, S_3$ | — |

Figure 6.16  Control unit for binary multiplication

|  | $Q_o = 0$ | $Q_o = 1$ |
|---|---|---|
| $q_0$ | $q_1, S_1$ | $q_1, S_1$ |
| $q_1$ | $q_2, S_3$ | $q_5, S_2$ |
| $q_2$ | $q_3, S_3$ | $q_6, S_2$ |
| $q_3$ | $q_4, S_3$ | $q_7, S_2$ |
| $q_4$ | $q_0, S_0$ | $q_0, S_0$ |
| $q_5$ | — | $q_2, S_3$ |
| $q_6$ | — | $q_3, S_3$ |
| $q_7$ | — | $q_4, S_3$ |

Figure 6.17  Alternative multiplication control unit

**Example 6.7**

We shall design a binary division circuit which implements the division algorithm specified in Procedure 2.8, using repeated subtraction. This algorithm which computes $X/Y$ where $X$ and $Y$ are integers can be specified in our register transfer language as follows:

| Division Algorithm | Explanation |
|---|---|
| $T_1: S_1: \quad B \leftarrow X; Q \leftarrow Y;$ <br> $R \leftarrow 0; A \leftarrow 0;$ <br> $D \leftarrow n$ | The two operands $X$ and $Y$ are stored in Registers $B$ and $Q$ respectively. Registers $A$ and $R$ are initialized to 0 and the number of bits, $n$, is stored in $D$ |
| $T_2: S_2: \quad A \leftarrow SL(A);$ <br> $B \leftarrow SL(B);$ <br> $A_0 \leftarrow B_{n-1}$ | The most significant bit of $X$ is shifted into $A_o$. |
| $T_3:$ IF $(A \geq Q)$ THEN $S_3:$ <br> $A \leftarrow A - Q; R_0 \leftarrow 1,$ <br> ELSE GO TO $T_4$ | If $A \geq Q$, then set result register bit $R_o = 1$, and subtract $Q$ from $A$. |
| $T_4: S_4: \quad A \leftarrow SL(A); B \leftarrow$ <br> $SL(B); A_o \leftarrow B_{n-1};$ <br> $D \leftarrow D - 1; R \leftarrow SL(R)$ | Shift next most significant bit of remainder of $X$ into $A_o$, decrement $D$, shift result register left. |
| $T_5:$ IF $(D = 0)$ THEN $S_o:$ read next instruction, ELSE GO TO $T_3$ | Repeat $n$ times until $D = 0$, at which time the integer part of $X/Y$ is in $R$ and the remainder is in $A$. |

The control unit is shown in Figure 6.18(a) and the basic system structure is shown in Figure 6.18(b). The subcircuits labelled $N_{AB}$, $N_D$, $N_R$, $N_Q$ are combinational circuits which generate the controlled excitations for the corresponding register flip-flops. The control unit state table can be reduced by combining the equivalent states $q_2$, $q_3$, and $q_4$. By initializing $D$ to $n + 1$ instead of $n$, $T_2$ is modified so that the operations executed during $T_2$ and $T_4$ are identical, and the control unit can be reduced to two states (Problem 6.5). Note that the division algorithm assumes the existence of subtraction and comparison logic in the CPU combinational circuit.

For each macroinstruction an algorithm of microinstructions must be selected to be implemented. The choice of this algorithm will affect the logical complexity of both the control unit and the CPU as well

|       | $D = 0$     | $D \neq 0$<br>$A \geq Q$ | $D \neq 0$<br>$A < Q$ |
|-------|-------------|--------------------------|-----------------------|
| $q_0$ | $q_1, S_1$  | $q_1, S_1$               | $q_1, S_1$            |
| $q_1$ | ——          | $q_2, S_2$               | $q_2, S_2$            |
| $q_2$ | ——          | $q_3, S_3$               | $q_4, S_4$            |
| $q_3$ | ——          | ——                       | $q_4, S_4$            |
| $q_4$ | $q_0, S_0$  | $q_3, S_3$               | $q_4, S_4$            |

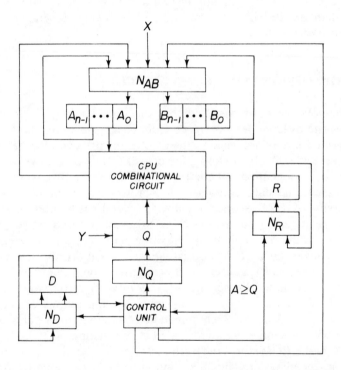

**Figure 6.18**    (a) Control unit state table and (b) System block diagram for binary division

as the speed of operation of the system. Thus in choosing an algorithm the designer must consider what microinstructions are required by other macroinstructions (to select a basic set of microinstructions) and the tradeoff between speed and complexity involved in implementing other microinstructions. A choice may also have to be made between implementing a given instruction as a macro or microinstruction. For instance if addition is a microinstruction, the CPU must contain a parallel adder. Alternatively addition can be implemented as a macroinstruction (Problem 6.6) in which case the CPU only needs a 1-bit adder as described in the table of Figure 4.12(a). This tradeoff is between increased speed and reduced control unit complexity (if implemented as a macroinstruction) and reduced CPU complexity (if implemented as a microinstruction). The considerations and ramifications of these tradeoffs may be quite complex. Consequently, digital system design is very much a heuristic rather than an algorithmic process, and is highly dependent on the experience of the designer.

## 6.5 LOGIC DESIGN USING *ROM's* AND *PLA's*

Up to now we have assumed that the combinational logic circuits and subsystems have been designed from basic gate elements. However in integrated circuit technology memory elements can also be utilized to realize logic. A Read-Only Memory (*ROM*) operates in a similar manner to a conventional memory unit in the sense that when a word is addressed, the data contained in that word is read out. However the contents of a *ROM*, once initialized, are not easily altered. Because they are relatively inexpensive, *ROM*'s have been utilized to perform logical functions. A combinational function $f(x_1, x_2, ..., x_n)$ of $n$ variables can be realized by a *ROM* consisting of $2^n$ 1-bit words. Each of the $2^n$ possible values of x, causes a distinct word of the *ROM* to be accessed and that word should be specified (*programmed*) as the value of $f$ for that particular input. Thus for any input $x_i$, the output of the *ROM* would be $f(x_i)$. A set of $k$ combinational functions of $n$ variables can be realized in a similar manner by a *ROM* consisting of $2^n$ $k$-bit words. Effectively the *ROM* stores the truth tables of the functions to be realized and upon receiving any input $x_i$, outputs the corresponding function values listed in the truth table.

It is possible to design complex combinational circuits from interconnections of *ROM* modules. Thus if a *ROM* with eight 2-bit words is programmed to realize a 1-bit adder module (as specified in the truth table of Figure 4.12(a)), a parallel adder can be designed from an interconnection of such *ROM*'s as shown in Figure 6.19.

For a set of $k$ $n$-variable functions several of the $2^n$ possible inputs

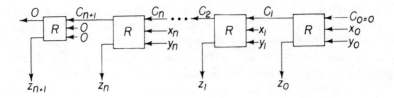

**Figure 6.19  ROM design of parallel adder**

$x_i$ may generate the same values of each of the $k$ functions. Thus for the one bit adder modules if $x_i = 0$, $y_i = 1$, $C_i = 0$ the outputs are $Z_i = 1$ and $C_{i+1} = 0$, and the same outputs are generated for the inputs $(x_i, y_i, C_i) = (0,0,1)$ or $(1,0,0)$  The question arises whether it is possible to take advantage of this fact to reduce the memory requirements of a memory logic realization, by having different inputs access the same word of the memory. A set of $k$ $n$-variable functions could then be realized with a memory unit of $r$ $k$-bit words where $r$ is the number of distinct outputs generated by the $k$ functions.

A further simplification can be achieved if we allow one input $x_i$ to access several words of memory and generate the output as the logical OR of the accessed words.

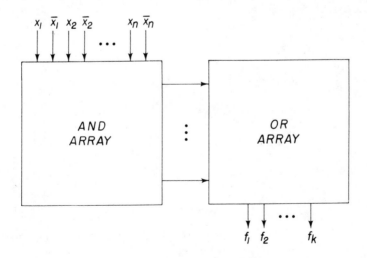

**Figure  6.20  Programmable logic array**

Both of these simplifications can be achieved by the use of *Programmable Logic Arrays (PLA)*. A *ROM* may be considered to consist of a decoder which translates an input into a memory address, and a memory unit to store data. The memory section of the *ROM* is programmable but the decoder section is fixed. A *PLA* differs from a *ROM* in that the decoder section can also be considered programmable. A *PLA* consists of a diode AND matrix (array) and a diode OR matrix as depicted in Figure 6.20. The AND matrix is programmed to generate product terms of the set of functions $F = \{f_1, \ldots, f_k\}$, and the OR matrix is used to OR together the appropriate product terms to generate the functions $f_i \in F$ as a sum of product terms. In effect the AND matrix

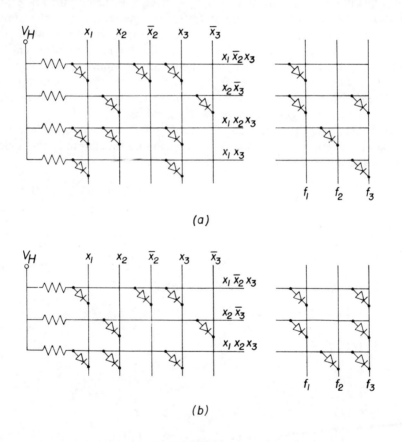

(a)

(b)

Figure 6.21   PLA design of a set of functions

is a programmable decoder and the OR matrix is a programmable data section. A set of functions can be directly implemented as a sum of product terms with sharing as depicted in Figure 6.21(a) for the set of functions, $f_1 = x_1 \bar{x}_2 x_3 + x_2 \bar{x}_3$, $f_2 = x_1 x_2 x_3$, $f_3 = x_1 x_3 + x_2 \bar{x}_3$. A more economical realization may sometimes be obtained as a multiple output minimal sum of products as shown in Figure 6.21(b). The control unit excitation logic may also be designed using ROM's or PLA's.

Since its inception the subject of switching theory has changed greatly, reflecting changes in the technologies which are prevalently used in the logical design of digital circuits. Undoubtedly further changes will be seen in the future in the technologies and design objectives of logical design. We believe this transition from one generation to another is simplified for the designer with a strong theoretical background. Our goal in this book has been twofold, first to present an introductory view of logical design and switching theory emphasizing problems and design techniques relevant to current technologies, and second, to develop a strong theoretical background on the part of the student so as to assist future transitions to new technologies. In the final chapter we shall consider some problems related to digital system design.

## SOURCES

The concepts of information transfer and register transfer languages were introduced by Bartee, Lebow and Reed[1]. Flores[4] presents detailed descriptions of the design of arithmetic logic networks.

A simple description of the design of a computer can be found in Booth[2], and Flores[5] and a more advanced treatment of this subject can be found in Chu[3]. Husson[6] is a good reference on microprogramming. Techniques for logic design using ROM's, PROM's (Programmable ROM's) and PLA's can be found in articles by Uimari[8] and Sypherd and Cassen[7] respectively.

## REFERENCES

[1] Bartee, T. C., Lebow, I. L. and I. S. Reed, *Theory and Design of Digital Machines*, McGraw-Hill, New York, N.Y., 1962.

[2] Booth, T. L., *Digital Networks and Computer Systems*, John Wiley and Sons, New York, N.Y., 1971.

[3] Chu, Y., *Computer Organization and Microprogramming*, Prentice-Hall, Englewood Cliffs, N.J., 1972.

[4] Flores, I., *The Logic of Computer Arithmetic*, Prentice-Hall, Englewood Cliffs, N.J., 1963.

[5] Flores, I., *Computer Design*, Prentice-Hall, Englewood Cliffs, N.J., 1967.

[6] Husson, S. S., *Microprogramming: Principles and Practices*, Prentice-Hall, Englewood Cliffs, N.J., 1970.

[7] Sypherd, A. D. and Q. C. Cassen, "System Design Concepts Using SOS Programmable Logic Arrays," *Proceedings of NEC*, vol. 30, 1974.

[8] Uimari, D., "PROM's—a practical alternative to random logic," *Electronic Products*, pp. 75-91, Jan. 21, 1974.

## PROBLEMS

**6.1)** Design a subsystem including the control unit to increment (if $I = 1$) or decrement (if $I = 0$) a 3-bit number by 1, assuming that logic exists to complement an individual bit.

**6.2)** Specify the control unit for an elevator for a three story building. The inputs are as follows:

$x_1$—first floor up button
$x_2$—second floor up button
$x_3$—second floor down button
$x_4$—third floor down button
$x_5$—elevator panel button for first floor
$x_6$—elevator panel button for second floor
$x_7$—elevator panel button for third floor

The two outputs are:

$z_1 = 1 \rightarrow$ elevator goes up
$z_2 = 1 \rightarrow$ elevator goes down
$z_1 = z_2 = 0 \rightarrow$ elevator stationary, stops

Requests are to be serviced in order of receipt with the exception of requests which can be serviced on route.

**6.3)** A machine is to be designed to add two floating point numbers $X$ and $Y$.

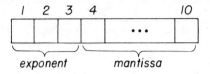

Figure 6.22    Problem 6.3 number representation

Bits 1-3 of each number are the 2's complement representation of the exponent and bits 4-10 are the 2's complement representation of the mantissa. The system works as follows. A pulse on $S$ initiates the addition of $X$ and $Y$ and the normalized sum $X + Y$ appears on outputs $Z$. A pulse on $D$ is to be generated at completion. If the result is too small or too large to be normalized with these bit fields a pulse is generated on output $U$ or $O$ respectively.

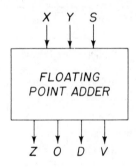

Figure 6.23    Problem 6.3 floating point adder

Specify a detailed design at the register transfer level assuming the operators, shift, increment by one, binary addition, and less than or equal to.

**6.4)** The following algorithm is used to convert a *BCD* number to binary.
  a) Shift the number right one position shifting in a 0 in the leftmost position.
  b) If the most significant bit of the $i^{th}$ digit is 1, subtract 3 from the $i^{th}$ digit for all $i$. If that bit is 0, do not subtract.

c) Repeat a), b). There are $4r$ iterations for an $r$-digit number. The numbers shifted out of the least significant position constitute the binary representation low order bits first

The following table illustrates the conversion of $110_{10}$ from *BCD* to binary.

|  | digit 2 | digit 1 | digit 0 |  |
|---|---|---|---|---|
| $110_{10}$ = | 0001 | 0001 | 0000 |  |
| shift | 0000 | 1000 | 1000 | $0 \leftarrow$ bit $a_0$ |
| (subtract 3 from digits 1 and 0) | 0000 | 0101 | 0101 |  |
| shift | 0000 | 0010 | 1010 | $1 \leftarrow$ bit $a_1$ |
| (subtract 3 from digit 0) | 0000 | 0010 | 0111 |  |
| shift | 0000 | 0001 | 0011 | $1 \leftarrow$ bit $a_2$ |
| shift | 0000 | 0000 | 1001 | $1 \leftarrow$ bit $a_3$ |
| (subtract 3 from digit 0) | 0000 | 0000 | 0110 |  |
| shift | 0000 | 0000 | 0011 | $0 \leftarrow$ bit $a_4$ |
| shift | 0000 | 0000 | 0001 | $1 \leftarrow$ bit $a_5$ |
| shift | 0000 | 0000 | 0000 | $1 \leftarrow$ bit $a_6$ |

Therefore $110_{10} = 1101110_2$.

Specify a system to implement this algorithm in RTL and derive a state table for the control unit assuming the existence of a shift register, a subtraction unit, and a logic circuit which can decrement by one.

**6.5)** Modify the binary division algorithm so as to make a two state control unit sufficient.

**6.6)** Specify an algorithm for binary addition in RTL assuming the only operators are logical AND, logical OR, and shifting. Derive the state table of a control unit for this algorithm.

**6.7)** Find an efficient *PLA* realization of the set of functions of Example 3.16.

# CHAPTER 7

# Physical Design and Testing

In the previous chapters we have considered many problems associated with the logical design of digital circuits and systems. In this chapter we shall briefly introduce two other problem areas of relevance to digital systems.

## 7.1 PHYSICAL DESIGN OF DIGITAL SYSTEMS

When the logical design has been completed, the *physical design* of a system must still be specified. The details of this problem will vary for different technologies. We shall consider the design problems relevant to *integrated circuits (IC's)* and *printed circuit cards*. The basic building blocks or elements in this technology are IC chips (or simply IC's). A single IC may contain several gates, or several memory elements, or even a complex functional unit such as a *ROM*. The independent subentities of an IC are called *portions*. The basic elements (gates, *FF*'s, *ROM*'s, *PLA*'s) of a digital system are each to be realized by an IC or a portion of an IC. Thus the first phase of the problem is to select a set of IC's which are sufficient to realize all elements of the system. (Note that a 3-input NAND can be used to realize a 2-input NAND by assigning the logical constant 1 to the extra input. Thus there need not be a strict equivalence between circuit elements and the IC portions used to realize those elements). The IC's are then placed on *cards*. In general one card cannot hold all IC's and hence several cards must be used. These cards are placed on a surface called the *backplane*. Electrical connections must then be made between IC portions. The three categories of connections are

(i) connections between two portions within an IC
(ii) connections between IC's on the same card
(iii) connections between IC's on different cards

We shall now consider in greater detail the sequence of problems which together constitute the physical design problem for integrated circuits.

(1) *IC Selection and Assignment Problems*—A set of IC's from a library of available IC's is selected to realize the circuit under consideration. The objective is to select a minimal cost set or a set containing the minimal number of IC's. The problem can be formulated in a manner similar to a covering problem, but the optimal solution may contain more than one of a particular IC type. Each circuit element is assigned to (i.e. realized by) a specific portion from the selected set of IC's. The objective is to minimize the number of inter-IC electrical connections required so as to simplify the subsequent interconnection problems.

(2) *Card Assignment Problem*—In general the complete set of IC's required to realize a circuit cannot be put on a single card. Therefore this set is partitioned into disjoint subsets, each such subset being assigned to a card. Electrical connections between cards or from inputs and outputs are made using card *connector pins*. The objective is to use as few cards as possible while not exceeding $P$ (a constant) connector pins or $M$ (a constant) IC's on any card. Thus in general the set of IC's should be partitioned so as to minimize inter-card electrical connections required, within the aforementioned constraints.

(3) *Card Placement Problem*—The IC's assigned to a given card are placed on that card, each IC being placed in one of $N$ prespecified positions. The objective is to simplify the routing of electrical connections between two IC's on that card and between IC's and connector pins for inter-card or input/output connections. The routing of these connections must frequently satisfy many constraints including a maximum length for any connection, a minimum distance separating two conducting paths, and no crossovers permitted between connections. The satisfaction of the latter constraint may require the card to have several *layers* upon which connections can be routed, with *passthroughs (vias)* between layers. It is impossible to state a simple objective function which precisely incorporates all of these constraints. A frequently used objective function for placement is

to minimize the sum of the lengths of all connections on the card. There is usually a good correlation between this approximate objective function and the difficulty of routing the connections on the card.

(4) *Card Interconnection*—The geometrical paths of all electrical connections are specified in accordance with the aforementioned constraint. Connections are usually routed one at a time. Associated subproblems include

   (a) *Ordering*—Which connections should be routed first? A frequently used rule is shortest connections first, but the effectiveness of this ordering is questionable

   (b) *Layering*—How many layers should be used and on which layer should each connection be routed?

   (c) *Routing*—Specify the geometrical path for a connection. The objective is to satisfy the constraints and not interfere with subsequent connections. The most frequently used procedure constructs a shortest path for the connection. This procedure which is due to Lee[6] and based on a maze-running algorithm of Moore[7] effectively finds a shortest path from point $A$ to point $B$ by exhaustive enumeration of all permitted paths from $A$ of length 1, 2, 3, . . . until point $B$ is reached.

(5) *Backplane Placement*—For circuits requiring more than one card, the cards must be placed on the backplane. The objective is to simplify the routing of inter-card electrical connections. This problem is similar to the card placement problem and its solution may utilize similar procedures and objective functions.

(6) *Backplane Interconnection*—The specification of the geometrical paths for all inter-card connections. This problem is similar to the card interconnection problem and may utilize similar solution procedures.

All of these problems are in general characterized by the fact that exhaustive evaluation of all possible solutions is infeasible. For instance if $M$ IC's are to be placed on a card with $N$ positions ($N \geqslant M$) there are $M!\binom{N}{M}$ possible placements. Exhaustive evaluation is only feasible for very small values of $N$ and $M$. Furthermore the objective of many of these problems cannot be formulated in a mathematically precise manner and therefore the objective function used is frequently only a goal approximation. Thus the true objective of a good card placement is to simplify the subsequent card routing, while the approximate objective function utilized is minimal total connection length.

For these reasons it is usually not attempted to find optimal solutions to these problems. Instead, procedures have been developed which are intended to produce relatively good solutions with respect to the approximate objective function in a computationally efficient manner. Such procedures are sometimes called *heuristics*. By the use of several different heuristics, or one heuristic several times, several solutions may be obtained which are relatively good with respect to the approximate objective function. These good solutions can then be evaluated in detail to determine the one which is best with respect to the *actual* objective.

## 7.2 TESTING AND DIAGNOSIS OF DIGITAL SYSTEMS

After a digital system has been completely designed, both physically and logically, it must be maintained during its period of service. This usually entails periodic testing of the system to determine possible malfunctions, and repair of the system when a malfunction is discovered. A *fault* which causes a circuit element to malfunction may produce an error at the circuit output. The process of testing consists of applying an input or input sequence to a digital system and concluding from the output response whether a fault exists (*fault detection*) and/or which element is faulty (*fault location or diagnosis*). In order to ensure the possibility of detecting existing faults by testing, it is usually assumed that faults are permanent (i.e. nontransient) and do not first occur during testing.

Consider an $n$-input combinational circuit which realizes the function $f(x_1, x_2, ..., x_n)$. If we assume that the circuit may fail in an arbitrary functional manner (i.e. the faulty circuit can realize any function $f'(x_1, x_2, ..., x_n)$) then in order to detect all such failures it is necessary to apply all $2^n$ inputs as tests. This is called *exhaustive testing*. Since this is infeasible except for small values of $n$, a more restricted class of possible faults is frequently assumed. The most commonly assumed class of faults is the *stuck type fault*, where some signal (or set of signals) in the circuit is transformed from a function of the inputs to a fixed logical constant. A signal which becomes fixed at logical one (or zero) is said to be stuck at one (or zero) which is denoted as $s$-$a$-1 (or $s$-$a$-0). In many technologies these correspond to the most common faults, shorts and open circuits.

Consider the circuit of Figure 7.1. The normal (i.e. fault free) circuit realizes the function

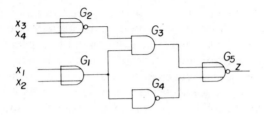

**Figure 7.1   Combinational circuit fault detection**

$$f = x_1 x_3 + x_1 x_4 + x_2 x_3 + x_2 x_4$$

The fault $\alpha$ which causes $G_2$ to become $s$-$a$-$0$ causes the output function to change to

$$f_\alpha = x_1 + x_2$$

The inputs which detect this fault are those for which the Boolean expression $f \oplus f_\alpha$ has the value 1, since for those inputs $f = 0$ and $f_\alpha = 1$ or $f = 1$ and $f_\alpha = 0$. In the illustrative circuit

$$f \oplus f_\alpha = (x_1 + x_2)\,\bar{x}_3\,\bar{x}_4.$$

Thus there are three tests which detect this fault, $(1,0,0,0)$, $(0,1,0,0)$, $(1,1,0,0)$.

An input which results in a different output for two different faults $\alpha$ and $\beta$, is said to *distinguish $\alpha$ from $\beta$*. The set of all such distinguishing tests consists of those inputs for which the Boolean expression

$$f_\alpha(x_1, x_2, \ldots, x_n) \oplus f_\beta(x_1, x_2 \ldots x_n)$$

has the value 1. Thus for the circuit of Figure 7.2

if                          $\alpha$ is the fault $x_2$ $s$-$a$-$1$

and                         $\beta$ is the fault $x_4$ $s$-$a$-$1$

then                        $f_\alpha = x_1 + x_3$

and                         $f_\beta = (x_2 + \bar{x}_3)\,(x_1 + x_3)$

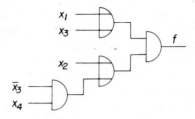

Figure 7.2   Combinational circuit fault diagnosis

and the set of tests which distinguish $\alpha$ and $\beta$ is defined by the expression

$$f_\alpha \oplus f_\beta = \bar{x}_2 x_3$$

and consists of the 4 tests defined by (-,0,1,-). Note that if $\gamma$ is the fault that the AND gate realizing $\bar{x}_3 x_4$ is $s$-$a$-1 then $f_\alpha \oplus f_\gamma = 0$. Thus no test exists which distinguishes $\alpha$ and $\gamma$ and these faults are said to be *indistinguishable*.

To simplify the computational aspects of generating a set of tests, it is frequently assumed that only a single signal in the circuit may be faulty, since a circuit with $s$ signal lines may have $2s$ single faults whereas it may have $3^s - 1$ multiple faults (exercise). The rationale behind the *single fault assumption* is that it is very improbable that several signals will fail simultaneously, and testing will be done sufficiently frequently so that a single fault will be detected and repaired before another fault can occur.

In some circuits it may be impossible to detect certain faults. For instance, in the circuit of Figure 7.3, if $\alpha$ is the fault that the AND gate which realizes the product term $x_2 x_3$ is $s$-$a$-0, then

$$f_\alpha = x_1 x_2 + \bar{x}_1 x_3$$

Figure 7.3   A redundant combinational circuit

and $f \oplus f_\alpha = 0$. Such a circuit is said to be *redundant*. Since some faults cannot be detected the single fault rationale becomes invalid for redundant circuits. However these circuits can be redesigned by removing gates and/or gate inputs so as to realize the same function in an *irredundant* manner so that all faults can be detected.

In testing a sequential circuit, an input *sequence* is applied to the circuit and the output *sequence* is observed. However the output sequence produced may vary dependent upon the initial state which is usually presumed to be unknown. Thus a fault $\alpha$ is detected by an input sequence I if and only if for all possible initial states, the output sequences produced by the normal circuit are different from those produced by the circuit with fault $\alpha$. Consider the *SR* flip-flop shown in Figure 7.4, and the fault *S s-a-0*. The input $S = 1$, $R = 0$ produces a 0-output if the initial state was $y = 0$ but it produces a 1-output if the initial state was $y$

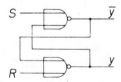

Figure 7.4    An *SR* flip-flop

$= 1$. Hence this input does not detect the fault since the normal circuit and the faulty circuit in initial state $y = 1$ produce the same output. However the input sequence consisting of $S = 0$, $R = 1$ as the first input followed by $S = 1$, $R = 0$ as the second input results in the second output $y = 1$ for the normal circuit and $y = 0$ for the faulty circuit independent of the initial state and hence detects the fault *S s-a-0*.

The problem of developing efficient algorithms to generate tests has received considerable attention. Such algorithms now exist for combinational circuits but efficient algorithms for sequential circuits have yet to be discovered.

A failure in the clock circuitry of a synchronous circuit may cause the circuit to exhibit asynchronous behavior. The study of asynchronous design techniques is important in analyzing these problems. In asynchronous circuits signal values which change concurrently, cannot be presumed to change simultaneously. This may create hazards as illustrated in Figure 5.11, which in turn may create errors in memory elements.

The subject of asynchronous circuit design develops procedures to circumvent such problems.

Although circuit faults can reasonably be assumed to be restricted to the stuck type variety, circuit design errors may not be so limited. The problem of *circuit verification* is a generalization of the fault testing problem in which it is attempted to distinguish a combinational or sequential function from all other such functions. For an $n$-state sequential machine $M$, a *checking sequence* is an input and output sequence which can be produced by $M$ but cannot be produced by any other $q$-state sequential machine where $q \leq n$. Any reduced strongly connected sequential machine has such a checking sequence, which serves to distinguish that machine from all other machines with the same number of states.

**Example 7.1**

Consider the following input and output sequence which is produced by a 2-state sequential machine.

Input Sequence:   0 0 1 0 0 1 0
Output Sequence: 0 0 0 1 1 0 0

Let us arbitrarily label the initial state $q_1$. Then $Z(q_1,0) = 0$. Now consider the state $q$ of the machine just before the fourth input. Since for the fourth input $Z(q,0) = 1$ then $q$ cannot be $q_1$ and hence must be $q_2$. Hence in response to a 0 input the machine produces a 0 output from $q_1$ and a 1 output from $q_2$. Thus the state after the first input is $q_1$. Therefore $N(q_1,0) = q_1$. Consequently the state after the second input is also $q_1$. Since the state after the third input was previously determined to be $q_2$, $N(q_1,1) = q_2$. Similarly since the state before and after the fourth input is $q_2$, $N(q_2,0) = q_2$. Finally the states before and after the sixth input must be $q_2$ and $q_1$ respectively so $N(q_2,1) = q_1$. Thus the only 2-state sequential machine which can generate this input and output sequence is defined by the state table of Figure 7.5

$$x$$

| | 0 | 1 |
|---|---|---|
| $q_1$ | $q_1,0$ | $q_2,0$ |
| $q_2$ | $q_2,1$ | $q_1,0$ |

Figure 7.5  State table derived from input output sequence

The problems of digital system design and digital system testing and verification are related since the difficulty of testing can frequently be reduced by clever design techniques. It is also possible to realize a function in such a manner that the circuit operates properly despite the presence of faults. For example, consider the realization of a function $f(\mathbf{x})$ which utilizes three sub-circuits with outputs $F_1(\mathbf{x})$, $F_2(\mathbf{x})$, and $F_3(\mathbf{x})$. If $F_1(\mathbf{x}) = F_3(\mathbf{x}) = F_3(\mathbf{x}) = f(\mathbf{x})$ and these three outputs are input to a logic module realizing the *majority function* $M(x,y,z) = xy + yz + xz$ the circuit output will be correct even if one of the three outputs $F_1$, $F_2$, $F_3$ is incorrect. This design procedure is called *triple modular redundancy (TMR)*. With the development of large scale integrated circuits, the difficulty of maintenance can be expected to increase, while the cost of individual components will probably decrease. Hence the reliability aspects of design will become an increasingly important consideration.

## SOURCES

Much work has been done on the two topics introduced in this chapter which are actually major areas of study in their own right. Many heuristic procedures which have been developed for the various physical design problems considered herein are presented in a book edited by Breuer[1], and are also presented by Friedman and Menon[5]. Likewise many procedures for the efficient generation of tests for digital circuits are presented in several books on this subject[2,3,4]. The predominant direction of research in both of these topics is to develop more efficient and fully automated algorithms to enable the solution of these problems for larger circuits and systems.

## REFERENCES

[1] Breuer, M. A. (editor), *Design Automation of Digital Systems, volume one, Theory and Techniques*, Prentice-Hall, Englewood Cliffs, N.J., 1972.

[2] Breuer, M. A. and A. D. Friedman, *Diagnosis and Reliable Design of Digital Systems*, Computer Science Press, Woodland Hills, Calif., 1975.

[3] Chang, H. Y., Manning, E. G., and G. Metze, *Fault Diagnosis in Digital Systems*, Wiley-Interscience, 1970.

[4] Friedman, A. D. and P. R. Menon, *Fault Detection in Digital Circuits*, Prentice-Hall, Englewood Cliffs, N.J., 1971.

[5] Friedman, A. D. and P. R. Menon, *Theory and Design of Switching Circuits*, Computer Science Press, Woodland Hills, Calif, 1975.

[6] Lee, C. Y., "An Algorithm For Path Connections and Its Applications," *IRE Trans. on Electronic Computers*, vol. EC-10, pp. 346–365, September, 1961.

[7] Moore, E. F., "Shortest Path Through a Maze," in *Annals of the Computation Laboratory of Harvard University*, Harvard University Press, Cambridge, Mass, vol. 30, pp. 285–292, 1959.

## PROBLEMS

**7.1)** In the circuit of Figure 7.6, which if any of the following tests detect the fault $x_1$ *s-a*-0?
(a) (0,0,0,0)
(b) (1,0,0,0)
(c) (1,1,0,1)
(d) (1,0,0,1)
(e) (1,1,1,1)

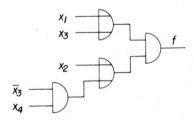

**Figure 7.6    Problem 7.1**

**7.2)** For the circuit of Figure 7.7 find a Boolean expression for the set of all tests which detect the fault

(a) $x_3$ s-a-0
(b) $x_2$ s-a-0
(c) $x_2$ s-a-1

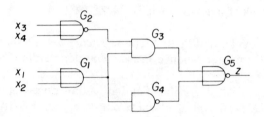

Figure 7.7   Problem 7.2

**7.3)** For the circuit of Figures 7.8(a) and (b) find a Boolean expression for the set of all tests which distinguish the following pairs of faults

(a) $x_1$ s-a-0 and $x_2$ s-a-1
(b) $x_2$ s-a-1 and $x_3$ s-a-0
(c) $x_1$ s-a-0 and $x_3$ s-a-0

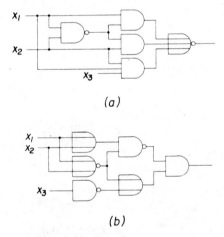

Figure 7.8   Problem 7.3

**7.4)** The following input-output sequence is generated by a two state sequential circuit with a single input and output. Find the circuit state table

Input Sequence:   0 0 1 0 0 1 0
Output Sequence: 0 0 0 1 1 0 0

**7.5)** A circuit has ten 2-input NAND gates and four 3-input NAND gates. There are two types of IC's . The first type has cost $k_1$ and contains four 3-input NAND gates. The second type has cost $k_2$ and contains two 3-input NAND gates and three 2-input NAND gates.

(a) If $k_1$ and $k_2$ are equal, find a minimal cost set of IC's to realize the circuit.

(b) For all possible relative values of $k_1$ and $k_2$ find the minimal cost set of IC's to realize the circuit.

**7.6)** The circuit of Figure 7.9 has six 2-input gates and is to be realized using an IC which has one 2-input gate which can realize an AND or NAND and one 2-input gate which can realize an OR or NOR. Three such IC's will be used.

(a) How many different IC assignments are there?

(b) What is the best such assignment.

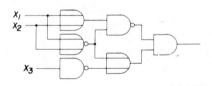

**Figure 7.9   Problem 7.6**

# Appendix

## Relay Contact Networks

A relay consists of an *electromagnet* which controls a set of *contact switches*. When the relay coil is energized, the contact switches change state, i.e. if they were normally open, they become closed and vice-versa. In a relay contact circuit the inputs are the signals applied to the coils. If a coil signal is $x_i$, then $x_i = 1$ if the coil is energized and $x_i = 0$ if the coil is not energized. A contact controlled by this coil will be closed if $x_i = 1$ and it is normally open, or if $x_i = 0$ and it is normally closed. Such contacts will be labelled $x_i$ and $\bar{x}_i$ respectively. A contact network consists of a set of interconnected contacts, which are controlled by a set of relays as shown in Figure A.1. An output $z_i$ takes the value 1 if and only if the relay excitation inputs are such as to cause a closed path in the contact network from $z_i$ to ground. Thus the function realized by the network of Figure A.2 is

$$z = x_1 x_8 + x_2 x_3 x_9 + x_4 x_5 (x_6 + x_7)(x_2 x_3 x_8 + x_1 x_9).$$

There is no direct (i.e. input for input) gate analogue of this circuit.

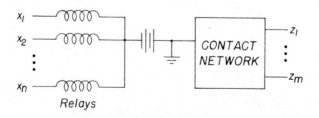

Figure A.1  A relay contact network

272

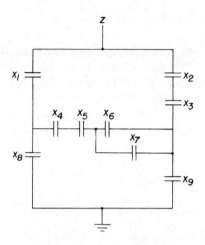

**Figure A.2  A contact network**

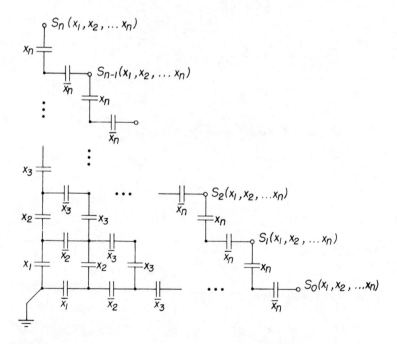

**Figure A.3  Contact network realization of symmetric functions**

## Realization of Symmetric Functions

A canonical network for realizing any symmetric function of the form $s_k(x_1,x_2, ...,x_n)$ is shown in Figure A.3. The circuit consists of contacts that may be opened or closed depending upon the value of the variable associated with it. A symmetric function of the form $S_{k_1,k_2, ...,k_p}(x_1,x_2, ...,x_n)$ is realized by connecting the outputs corresponding to $S_{k_1},S_{k_2}, ...,S_{k_p}$. Figure A.4 shows a realization of $S_{1,2}(x_1,x_2,x_3)$.

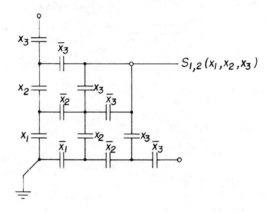

**Figure A.4   Contact network realization of $S_{1,2}(x_1,x_2,x_3)$**

# Index

275

278 Index